FUTURIST SPACEFLIGHT MEDITATIONS

Giulio Prisco

Giulio Prisco

CONTENTS

Title Page	
Futurist spaceflight meditations	1
Dedication	2
Acknowledgments	3
1 - Introduction	6
2 - Political preamble	14
3 - Lockdown	23
4 - Spaceflight now	32
5 - Spaceship Earth	42
6 - Cultural reflections	52
7 - Interplanetary	62
8 - Interstellar	71
9 - Cosmic engineers	81
References	91

FUTURIST SPACEFLIGHT MEDITATIONS

Giulio Prisco

Copyright 2021 Giulio Prisco

Cover picture: Anonimo caminante, by Franc Xavier Sánchez Jalao (Instagram: francxaviersan). "Anonimo caminante, tu que recorres la senda de un destino incierto roto por la grieta del tiempo, difumina tus pensamientos a la espera del casual reencuentro." Used with permission.

DEDICATION

To my stars: my beautiful wife and daughter Anna Maria and Melinda.

To my parents Anna and Glauco.

To my parents in law Magda and Vilmos.

To my grandparents Gigia, Giulio, and Rita.
Grandpa Amato, I didn't forget you, but we never met! I look forward to meeting you beyond the stars.

To Angelina, Enrica and Giovannella.

To my beloved pets Minou, Ricky, Sacha, and Speedygonzales.

To all my relatives and friends in the "real world." I wrote this book for you too, I hope you'll enjoy it!

ACKNOWLEDGMENTS

I wish to thank the authors and thinkers mentioned in this book, including those I disagree with. Of course, any misunderstandings and misrepresentations of their work are entirely my fault.

I wish to thank the readers of my previous related writings, and the early readers of the draft versions of this book, who sent me useful comments, suggestions, and objections: Adriano Autino, William Sims Bainbridge, Tom Bell, Howard Bloom, Steve Bowers, Riccardo Campa, Lincoln Cannon, Peter Garretson, Al Globus, Mikkel Haaheim, Alan Kazlev, Yalda Mousavinia, Philippe van Nedervelde, Catherine Newell, Ruben Novoa, Geoffrey Quick, David Roman, Gabriel Rothblatt, Martine Rothblatt, Rick Tumlinson, Adrian Tymes, Stefano Vaj, Arthur Woods. In particular, I wish to thank Alan and Geoffrey.

A problem with these lists is that one always forgets to include somebody who should really be in the list. If I have forgotten you, please let me know and accept my apologies! I also wish to thank the early readers who sent me anonymous observations.

I have listened carefully to all readers, but at times I have chosen to ignore some objections. Therefore, if you find in this book something that you really object to, please blame only me.

I wish to thank Franc Xavier Sánchez Jalao (Instagram: francxaviersan) for allowing me to use the cover picture. In the picture I see a stylized human walking in a starfield and spreading into the universe, which is what this book is all about.

The study and research that I did is not of the kind where one studies carefully a handful of books and papers. It's more of the kind where one goes rapidly through hundreds of books and papers, trying to learn from different angles.

I started practicing this learning style as a college student, and then as a young scientist working in research institutions with well stocked libraries. I used to take ten or twenty books and journals on related topics to a reading table, and go through them in parallel.

Today, I couldn't do this because I don't have the money to buy hundreds of books and research papers, many of which are outrageously priced. While I always buy the fairly priced works of independent authors, I am unable and unwilling to enrich major publishers.

However I do have the world's largest pirate research libraries at my fingertips, and everything is free. From Libgen and Z-Library, I can download most of the books I need. From Sci-Hub, I can download most of the research papers I need. Therefore, I wish to thank the Libgen, Z-Library, and Sci-Hub teams, and in particular Alexandra Elbakyan, the creator of Sci-Hub.

Yes, I know that piracy is bad. But I also know that knowledge and ideas want and need to be free. I'm an average sixty-something, but brilliant young people all over the planet also have the world's knowledge and best ideas at their fingertips. Some of them will make the world a better place. This is, I think, more important than the interests of the publishing industry.

Someday soon, I hope, there will be ways to fairly reward creators and at the same time make their creations affordable to everyone. In the meantime, since I enjoy the work of others without paying, I must make my own work freely available. The content of the free version of this book is identical to the retail version, and the retail Kindle version is not encrypted.

If you are reading the free version of my book, or a pirated copy of the retail version, I wish to thank you for your time and attention.

If you are one of those who bought the retail version, I wish to thank you for also buying me a coffee and some cigarettes.

1 - INTRODUCTION

We must strenuously push toward our cosmic destiny among the stars. Beginning to expand beyond the Earth before it's too late is our most important task at this moment in history. Many actors have important roles to play, and there's room for everyone. Spaceflight will also help find viable solutions for current developmental, environmental, and social problems.

But the road to the stars is full of impediments and roadblocks. We will not advance as fast as we wish. Therefore we must keep our mood strenuous and our drive strong. We need an optimistic spaceflight culture oriented to the future, with energizing visions of interplanetary, interstellar, and cosmic futures. We also need a futurist space philosophy.

My name is Giulio Prisco. I was born in 1957. On July 20 and 21, 1969, I was glued to a black and white TV watching Neil Armstrong and Buzz Aldrin walk on the Moon.

I grew up reading a lot of science, science fiction and space stuff. I studied theoretical physics; moved to CERN; then on to the European Space Agency (ESA); and from there, to senior management roles in other public space agencies.

In 2005 I left public service to start a technology consulting company that was reasonably successful for a few years until the recession hit. Then I became a free agent dabbling in many different areas, including business and scientific journalism for a living, and philosophy for fun. Besides spaceflight, I'm especially interested in the intersections of science, metaphysics, and religion ([Chapter 9], [Prisco 2020]).

One thing has been a constant in my life: after my family, spaceflight is the most important thing in the world to me.

I have much more than space exploration in mind. Not merely space exploration, but space expansion. I want to see humanity expand beyond the Earth and establish human societies and industries in outer space. Following Adriano Autino [Autino 2020], I support "the growth and expansion of civilization into space as a strategic long-term philosophical objective" that also needs immediate action.

I want to see people in space. Their nationality, their political ideology, the political ideology of their nation, what they do in space, their skin color, gender, sexual preferences, religion... I couldn't care less. I just want to see people - human beings like you and me - opening the space frontier.

Spaceflight is our existential imperative, built into our human nature. "I think we're going to the Moon because it's in the nature of the human being," said Neil Armstrong, the first man to walk on the Moon, as reported by Norman Mailer [Mailer 1970]. "It's by the nature of his deep inner soul. Yes, we're required to do these things just as salmon swim upstream."

Going to the Moon and to the planets, said John Glenn, the first American astronaut to orbit the Earth, as reported by Oriana Fallaci ([Fallaci 2014], [Chapter 6]), "isn't a question of right: it's a duty."

This is it. Our duty to God, or to God by any other name, or to the cosmos, or to some cosmic principle that favors life, or to life itself, is to expand beyond the Earth into the black sky.

Frank White extends James Lovelock's concept of Gaia, the living Earth ([Chapter 5], [Lovelock 2000]), to the whole universe. The universe itself will become a living whole and we are "actively encouraged by larger forces" [White 2021] to expand beyond the Earth and "help the universe become increasingly self-aware" [White 2018].

I think our built-in drive to explore and expand into outer space reflects the wonderful, transcendent cosmic destiny [Chapter 9] that is waiting for us out there among the stars. To me and others like me, this gives meaning and urgency to our first steps, here and now, on the road to the stars. This, not the interests of this or that nation or corporation, is why humanity must strive to become interplanetary [Chapter 7] and then interstellar [Chapter 8].

Of course I realize that others don't necessarily share my philosophical convictions. We can only win their support for spaceflight based on more practical considerations. I'll try to provide the best practical arguments for spaceflight that those who don't hear (or don't want to hear) the call of the stars can also relate to.

Philosophical convictions aside, there's a strong ultimate argument to show that humanity must spread to the stars: one day our Sun will die, and we'll have to live around other stars. But the Sun will die in the very far future, so this is unrelated to the choices that we have to make here and now [Deudney 2020].

Another strong, but untimely, argument for spaceflight is that mass migrations to space would reduce the human population of the Earth, and reduce human pressure on the biosphere. This is true, and hopefully we'll have the option to move to, say, a terraformed Mars or engineered space habitats [Chapter 7], in a few centuries or so. But we must do other things first to prepare the way.

We must begin to expand beyond the Earth before it's too late. An asteroid impact could wipe out human civilization anytime. We could wipe out human civilization ourselves in war. New pandemics of natural or artificial origin could end us. And if we let ourselves fall into cultural senility, we could be unable to rise up again.

Of course I have my own preferences. I would like to see humanity expanding into space in certain ways instead of certain other ways. But space expansion is too important for yes-buts. I don't say yes, but only if this or that. I say yes, period.

For example, I don't like war but I'll welcome a military push to space if nobody else is pushing. I'm for decentralization and small governments and businesses, but I'll welcome big central governments and/or big businesses to lead the way to space. I don't like the totalitarian ways of contemporary China, but I'll welcome China to lead the way to space if the West doesn't. I'm not overly fond of some trends in contemporary Western culture (e.g. forced political correctness), but I hope those same trends can be leveraged to win more support for spaceflight in the West.

I'm one of many orphans of Apollo. We were given a beautiful dream when we were kids. Then the dream was taken away from us. The Apollo program was not designed as a sustainable spaceflight program, and therefore it was doomed from the beginning.

"As William Sims Bainbridge pointed out in his 1976 book [Bainbridge 1976], space travel is a technological mutation that should not really have arrived until the 21st century," said Arthur Clarke [Bainbridge 2007]. The Moon "was reached half a century ahead of time."

The Apollo adventure of the 1960s was but an inspiring prelude of things to come. Our interplanetary and interstellar future is the end point of the ongoing process that was initiated by the Apollo program. We are still confined to our planet, but we are striving to develop scientific, technical, economic, cultural, and political enablers for the real beginning of humanity's expansion into space.

We are part of this process as crew members of Spaceship Earth [Fuller 1969], en route in its mission to the universe, and what we do matters. Besides piloting Spaceship Earth and tending to the propulsion engines, we must take care of the life support system and the well-being of the crew [Fuller 1969, Wolfe 2015]. Spaceflight enthusiasts should see these tasks - protecting the Earth and other people - as integral parts of our common space mission [Chapter 5].

We Apollo orphans shouldn't think of the post-Apollo decades as time wasted, but as time spent developing much needed technologies, infrastructure, and other enablers (including cultural enablers) for new, sustainable and permanent waves of human expansion into space.

Now we have much better computer and telecom technology than in the Apollo years. "Much better" is an understatement: "spectacularly better" is more like it. Computing and telecom have been the highest growth industries in the post-Apollo decades. Now we can watch spaceflight as it happens on YouTube in HD, with stupendous detail. Tomorrow, Virtual Reality (VR) will allow us to walk on the Moon and Mars and talk to those who are physically there.

The Space Age that we imagined in the Apollo years has morphed in a very real "unexpected space age," notes Sarah Cruddas [Cruddas 2020]. The blending of satellites, computing and telecom has created the smartphone based location services that we find indispensable today. Next time you use Waze, think of all the satellites in space that ensure you are *"Never Lost Again"* [Kilday 2018].

It's worth noting that many Apollo orphans have played important roles in the development of today's computing and telecom technology, which is a necessary enabler of long-range space exploration and expansion.

Biotechnology [Mason 2021], automation, robotics, VR, additive manufacturing (3D printing) and eventually nanotechnology [Drexler 1986, Bainbridge 2007, Prisco 2020], narrow Artificial Intelligence (AI) and eventually real AI [Bainbridge 2007, Prisco 2020], will also play a growing and very important role out there.

Looking back, one can have the impression that in the Apollo years everyone in the US was in love with spaceflight and willing to support as much public funding as needed to reach the Moon and then beyond. But it was not so. In *"The Meaning and Value of Spaceflight"* [Bainbridge 2015], William Sims Bainbridge notes that "at no point... did a majority of the American public support increased funding for space exploration." But "there was sufficient public support so that political elites could invest in the program." In other words, in the Apollo years public support for spaceflight was slightly above critical mass in the US.

Bainbridge's excellent book, which I'll frequently cite, analyzes public perceptions of spaceflight since the second world war and correlates support for spaceflight with demographic variables including wealth, race, gender, and education.

Writing before 2015, Bainbridge emphasized that he didn't see "any evidence that the general public has turned its back on spaceflight." I think the difference between today and the Apollo years is that public support for spaceflight in the US, which was slightly above critical mass then, is slightly below critical mass now. If so, American spaceflight enthusiasts only need to win a few percent points of public support, which seems attainable.

Bainbridge's analysis reveals that a majority of the American public, now like in the Apollo years, would support spaceflight if only it didn't cost so much. In fact, the most frequently heard argument against space exploration is that the money would be better used for more important things here on Earth.

A good counter argument is that space exploration costs much less than the money that we consumers spend on useless things, and much less than other questionable public expenses. It can be argued that ambitious Apollo-like space programs could be entirely funded with the money that are now wasted by government programs (including waste in space programs themselves).

Spaceflight doesn't have to cost as much money as in the Apollo years. According to Bainbridge, "a renaissance of space exploration could come about after other fields of science and technology had independently established the preconditions for new and more efficient launch systems." A spectacular reduction in the cost of launch systems and space operations is ongoing today [Chapter 4].

This book goes from the here and now to the far future and the stars. I'll start with current issues here on Earth: politics and culture wars [Chapter 2], and the disastrous potential effects of a planetary lockdown [Chapter 3].

In [Chapter 4] I'll provide a high level bird's eye view of spaceflight today and in the next few decades. There are promising indications that we could return to the Moon permanently and sustainably, and then move on to Mars and the asteroids. Of course, what will happen in the next few decades strongly depends on geopolitical issues here on Earth, and internal political issues in the main spacefaring nations. But also the private sector has a key role.

There are many actors on the stage [Goswami 2020]: big governments, small governments, large corporations, lean startups, more and more citizen organizations and interest groups. Different actors push for different ways of doing things, e.g. national or global, government or private sector, civilian or military, humans or robots, Moon or Mars. I think space is a big place, and there's room for many actors and a wide range of approaches.

In [Chapter 5] I'll discuss spaceflight in the context of developmental, environmental, and social issues. A main point is that the drive toward spaceflight doesn't exclude, but requires, making the Earth a better place. Conversely, spaceflight will result in a wealthier and greener Earth, and a nicer society.

Besides practical factors (money, politics, technology) I'll emphasize the key importance of cultural factors in [Chapter 6]. I'll combine a necessarily pragmatic approach to actual spaceflight here and now with great expectations for our future in space, and also emphasize the need for a futurist philosophy of spaceflight.

I realize that spaceflight initiatives must be realistic, pragmatic, and sustainable. This can only be a slow hike, not a sprint. But we need to establish, urgently, cultural and philosophical movements able to sustain our long, slow hike to the stars.

In the last three chapters I'll outline optimistic, energizing visions of interplanetary, interstellar, and eventually cosmic futures. These visions can provide inspiration and motivation for our work here and now.

I'll present arguments, strong arguments I think, that could persuade readers whose primary interest is national security, or patriotism, or the economy, or climate change, or social justice, or whatever, to support spaceflight. But I won't hide that my own primary interest is spaceflight itself, and some of those readers are likely to find this annoying, like "this space cadet only wants to keep his beloved rockets flying."

So I have written this book for space enthusiasts like me. If you are one of us, I hope you will find in this book good arguments to confirm your enthusiasm for spaceflight, and good arguments that you can use to persuade others. My hope is that you will use these arguments to speak up, loud, clear, and very persuasively.

I want this to be a short book, so I'll touch on many things without going in depth. However, I'll try to provide the best, more informative and insightful external references.

Following the Chinese convention, I'll refer to Chinese nationals with family names first. Literal quotations are enclosed by quotation marks ("like this"). Both internal and external references are indicated in square brackets (e.g. [Chapter 1], [Prisco 2020]). I won't use footnotes or endnotes.

2 - POLITICAL PREAMBLE

These days I tend to avoid discussing politics and culture wars [Lakoff 2016], especially on social media. Unfortunately, today these discussions inevitably degenerate into name calling, and I have better things to do.

But spaceflight is very much conditioned by the political and cultural climate, and therefore much of this book is about politics and culture. So I must disclose where I am coming from, and I'd better do so right here at the beginning.

When I was 13 or so I read somewhere that the US would likely abandon the Apollo program and the Moon. I was devastated, because spaceflight was the most important thing in the world to me.

But the writer also said that then the communist Soviet Union would likely go ahead with its own Moon exploration program and take the lead. To me, the Moon was more important than politics here on Earth. So I decided to become a communist.

I studied communism intensely. I thought I would love a society of equals where everyone strived selflessly toward common goals, but something told me that the real world is more complicated than books.

I think healthy societies could be based on different economic and political systems, including capitalism and communism. In theory. But in practice, "Under capitalism, man exploits man. Under communism, it's just the opposite." My interpretation of this quote (often attributed to John Kenneth Galbraith) is that many economic and political systems, which could work in theory, don't work in practice due to our human nature.

If only we weren't such greedy, power hungry assholes, the world would be a much better place. But human nature is a given, and we must work with people as they are.

I assume that all readers are familiar with the terms "liberal," "conservative," and "libertarian" as used in current political discourse. I'll just note that today's liberalism doesn't have much to do with classical liberalism, which in contemporary terms would be described as moderate conservatism leaning toward libertarianism [Lakoff 2016].

I prefer not to describe myself as a conservative or a liberal. I agree with some liberal ideas, and I agree with some conservative ideas, and I dislike some toxic excesses on both sides, and that's it. Of course, many liberals call me a conservative, and many conservatives call me a liberal, and often with personal insults, but believe me I couldn't care less.

I'm one of those "biconceptuals" that, according to (liberal) thinker George Lakoff [Lakoff 2014, 2016], can relate to both liberal and conservative conceptual frameworks and value systems. We are swing voters. In the US, swing voters are a quarter of the total but, since pure conservatives and liberals are split more or less equally, swing voters often get to decide elections.

According to Lakoff, there's no such thing as a moderate: one can be a moderate liberal, or a moderate conservative. Even biconceptuals lean one way with "one worldview more strongly held or more widely applicable than the other." However, I think a moderate conservative and a moderate liberal have much in common and can work together to achieve practical outcomes that both want.

Lakoff reports (but disagrees) that libertarians "see themselves as forming a separate political category, neither liberal nor conservative, but something unto itself." I look very much forward to seeing libertarianism grow, but I realize that current politics is more and more polarized between the two main camps, and this is a fact we have to live with.

I'm a libertarian at heart, but I can be persuaded to pragmatically support non-libertarian policies, and even mildly authoritarian policies as temporary least-evil measures.

According to Lakoff, "a libertarian is two steps away from a mainline conservative" [Lakoff 2016]. I disagree, because according to Lakoff's own framework a libertarian can also be seen as two steps away from a mainline liberal. I guess Lakoff dislikes both conservatives and libertarians, and so he wants to put them together in one dislikable category. Lakoff goes on to define a libertarian as "an extremely pragmatic conservative whose moral focus is on noninterference by the government."

For me, noninterference by the government is a focus but not the only one. I'm definitely pragmatic though: I don't care for ideological purity, to which I prefer negotiated solutions that work.

Most people, I think, sincerely want to make the world a better place for everyone. But this doesn't imply that they always agree on the best path to get there from here. I think it's perfectly normal, and inevitable, that we have different interests and priorities. Well-meaning people can disagree, for reasons that are perfectly valid to them. Therefore, there'll always be tensions and conflicts, and we'll always need negotiation. Politics is, or should be, the art (art, not science) to sit down and negotiate some way forward that everyone around the table is willing to live with for a while.

I'm not against governments. I often find bureaucrats annoying, but I realize that they do important work that I would be unable to do. I have been a bureaucrat so I learned the skill set, but I never had the slightest inclination for the work. But it is important work that somebody must do.

I'm hardly the only libertarian at heart who recognizes the need for good government. Stewart Brand of *Whole Earth Catalog* fame describes himself as a post-libertarian, praising the government and the excellent work done "by people with narrow ties who worked nine to five, often with federal money" [Wiener 2018].

In the words of moderate conservative Ross Douthat [Douthat 2020] "centrists and self-consciously bipartisan observers" like me yearn "for a lost age of compromise" and condemn "the ideological polarization of the parties, and the disastrous effects of polarization." In my case, this is very much correct.

I'll try to provide arguments for spaceflight that both conservatives and liberals can relate to. Of course, some liberals will see this book as conservative propaganda, and some conservatives will see it as liberal propaganda. But I see it as a nonpartisan, or bipartisan, or biconceptual, or third way framework that people with different ideological leanings can support without betraying their favorite ideology.

I'll borrow the words of Bill Gates, who hopes we can unite behind plans that bridge political divides. "Whether you're a believer in the private sector, or government intervention, or activism, or some combination, there's a practical idea you can get behind," he says in his climate change book [Gates 2021]. "As for the ideas you can't support, you may feel compelled to speak out, and that's understandable. But I hope you'll spend more time and energy supporting whatever you're in favor of than opposing whatever you're against."

It seems that, today, conservatives tend to support spaceflight more than liberals. But this hasn't always been the case, and doesn't necessarily have to be the case. Plenty of liberals have supported and continue to support spaceflight. The Apollo program was started by Democratic President John Kennedy, and killed by Republican President Richard Nixon [Launius 2019].

"Ecology and technology find a unity in Space," said legendary liberal politician Jerry Brown [Brand 1977]. "Going into Space is an investment. It's not a waste of money, it's not a depleting asset, it's an expanding asset, and through the creation of new wealth we make possible the redistribution of more wealth to those who don't have it."

"As long as there is a safety valve of unexplored frontiers, then the creative, the aggressive, the exploitive urges of human beings can be channeled into long term possibilities and benefits," added Brown. "But if those frontiers close down and people begin to turn in upon themselves, that jeopardizes the democratic fabric."

The potential of spaceflight to make the world wealthier, greener, and fairer, and the importance of an open space frontier, are explored in the following chapters. I think we should return to the Moon first, sustainably and permanently, and then move outward.

In the post-Apollo US, new administrations have often abandoned space programs launched by previous administrations. This is a big problem because, like in most democracies, the US administration typically changes color every eight years, or even four, which is not enough for a new space program to take off.

Therefore, only bipartisan support can lead to stable space programs. Unfortunately, today the gap between two political camps that hate each other, and are unwilling to negotiate bipartisan agreements for the common good, seems wider than ever in the US and other Western democracies. This trend, which is likely to continue and even grow stronger, makes achieving bipartisan support for space programs difficult.

But the world is bigger than the US and the West. In particular, China is rising and challenging the supremacy of the US [Kissinger 2011, Mahbubani 2020].

In passing, wouldn't it be wonderful if there were no nations but one united humanity under one world government? Yes, perhaps it would. But only if the world government is a good government. A brutally totalitarian world government would be a nightmare, if you ask me, from which nobody can escape.

Please don't tell me that can't happen, because history shows that good can turn to bad very fast. Perhaps one day humanity will be united, but at this moment I like the option to move elsewhere if things turn bad. Similarly I like that, if one nation stops leading spaceflight, other nations will take the lead.

It's worth noting that "the China dream," a favorite slogan of China's leader Xi Jinping, is also the title of a book written by military expert Liu Mingfu [Liu 2015]. The book, first published in Chinese in 2010, was a much discussed best seller in China. The author envisions China's supremacy, enabled not only by an economic rise but also by a military rise and a "martial spirit."

China wants to rise to space as well and is advancing, step by step, toward achieving supremacy in the cislunar space around the Earth and the Moon ([Chapter 4], [Goswami 2020]).

China is not a democracy in the Western sense, and therefore the Chinese space program is more stable than the US space program, with less internal political roadblocks and more ability to execute long-term plans. It seems that China is advancing slowly but steadily like a glacier, and the West is stumbling around like a drunkard. This exaggeration has, I think, elements of truth.

American politicians and citizens should realize that Chinese supremacy in cislunar space (or, using the title of Kim Stanley Robinson's very relevant science fiction novel, a "Red Moon" [Robinson 2018]) and beyond could have disastrous consequences for the US.

A Red Moon would inevitably result in important economic, military, geopolitical and cultural advantages for China. The rest of the world would look to China as the promised land of unlimited progress. Just like, you know, the rest of the world used to look at America.

I'm not an American, and so the prospect of a Red Moon is not that disturbing to me. My first reaction is, if it is inevitable that China takes the lead, so be it, and here's to China. To make peace with this, I've been reading a lot about China's past and present, and a lot of Chinese literature in translation.

So my state of mind is similar to that of my young self in the Apollo years. Should I become a fan of communist China? Or, more precisely, a fan of today's China of Xi Jinping [Brown 2016], whose ideology is quite different from traditional communism?

But I'm a libertarian at heart, and therefore I can't be a fan of today's China, which is an authoritarian surveillance state where people are invasively monitored from the cradle to the grave [Strittmatter 2019] and punished for all sorts of noncompliant behaviors and opinions.

So on the one hand, since spaceflight is of overwhelming importance to me, I'm willing to make peace with the idea that China will lead the way to the stars in the rest of this century. But on the other hand, I prefer the Western culture in which I've been raised, and I hope we'll continue to be among the leaders.

Kishore Mahbubani notes [Mahbubani 2020] that, while Chinese culture "values social harmony over individual empowerment," American culture is the opposite, and this is an important strategic advantage of America over China. Americans argue "loudly and vociferously over the direction that America should take." Open and vigorous dissent is one of the factors that made America strong, and America needs open and vigorous dissent to keep strong. I totally agree.

This brings me to forced political correctness, identity politics [Chua 2018], and "woke" liberalism, described by Douthat as an ideological movement characterized by "airless certainties and willful disdain for moderation." Meghan Daum, who identifies as a liberal, describes woke liberalism as "leftist sanctimony run amok" [Daum 2019].

I find some excesses of woke liberals very annoying, but I understand that they are often motivated by sincere outrage at the lack of social justice in our past and present, and I can empathize with those who want revenge. However, history shows that, rather than ending oppression, often the oppressed of yesterday and today want to become the oppressors of tomorrow. While this is a perfectly understandable human reaction, I think it's dangerous and counterproductive.

I hope that the current outbreak of "cancel culture" [Daum 2019] in the West is only a temporary phenomenon. Otherwise, China will win. In passing, it's worth noting that today's Western cancel culture is similar in spirit to the Chinese cultural revolution launched by Mao Zedong in the sixties.

Mao unleashed "youth militias bonded by ideological fervor" to cancel "old ideas, old culture, old customs, and old habits" [Kissinger 2011]. Dissenters were publicly shamed, humiliated, severely punished, and forced to ideological reeducation (or worse). The militias also went to war with each other. The parallels with today's cancel culture in the West are evident, and indicate that we are on a very dangerous slippery slope. Eventually, Mao had to call in the army to stop the violence of the precursors of today's cancel mobs.

Amy Chua notes [Chua 2018] that, at different times in the past, "both the American Left and the American Right have stood for group-transcending values. Neither does today." I think we need to protect group-transcending values, of which spaceflight is a good example, from today's identity politics and cancel culture.

Chua emphasizes that identity politics "inevitably subdivides, giving rise to ever-proliferating group identities demanding recognition... dividing people into ever more specific subgroups created by overlapping racial, ethnic, gender, and sexual orientation categories." Often these subgroups fight each other like Mao's militias (fortunately with tweets instead of guns), which makes me hopeful that cancel culture will cancel itself one day.

I hope that one day, hopefully soon, conservatives and liberals will let go of the toxic aspects of their respective ideologies, let go of their hatred for each other, and sit the fuck down to discuss and negotiate. Perhaps this is even less likely to happen soon than the faster than light interstellar travel schemes outlined later in this book, but I prefer to be optimistic.

I hope the West will continue to play important roles in our expansion into the solar system and beyond. This book is, among other things, a call for the West to react.

3 - LOCKDOWN

I started writing this book in 2020. The year will be remembered for the first launch of NASA astronauts to the International Space Station with reusable rockets designed and built by a commercial company, Elon Musk's SpaceX.

This was very significant because it combined two factors that are key to sustainable spaceflight: the direct involvement of the private industry, and the launch cost reduction enabled by reusable rockets. It's expected that these two factors will continue to bring down the cost of spaceflight.

But 2020 will also be remembered for the COVID-19 pandemic and the resulting lockdown. All over the world, people have been locked down at home, and the lockdown continues in 2021.

The book *"Dark Skies"* [Deudney 2020], by Daniel Deudney, was published in 2020.

The central thesis of Deudney is that "we should fully relinquish the quest" for space expansion. Did you say "Bullshit?" I hope so.

Make no mistake: Deudney's book is well researched and written, and the author is a known and respected political scientist who understands space technology, politics, finance, and the impulse to expand into space. While Deudney doesn't like space expansionists like me, he understands us. In particular, he understands (but disapproves of) the "religion of cosmism" ([Chapter 6, 9], [Prisco 2020]) that is "at the core of habitat space expansionism."

Besides space expansionism, Deudney is against libertarianism, "the political philosophy of the politically naive (or the cynical rich)."

Some libertarian spaceflight enthusiasts want to go to space to escape regulations, but Deudney wants to close the space frontier to protect regulations, which are "likely to be reversed as humanity expands to other worlds." In space, regulatory interventions like "reversals and relinquishments will become vastly harder to achieve and maintain."

Of course, Deudney is against those who "expect radically improved humanity will conduct large-scale space expansion," and against the transformative technologies that could "radically improve the feasibility of many ambitious space projects as well as to drastically alter terrestrial conditions: genetic engineering, nanotechnology, and robotics and artificial intelligence."

Deudney understands that spaceflight "may be at the cusp of another major boom." Optimistic space expansion advocates "have been far more influential than their limited numbers might suggest." Transformative technology advances "appear very likely and are very likely to make space expansion considerably more feasible."

> "It seems reasonable to expect that over the course of this century options for conducting space projects will expand considerably, presenting humanity for the first time with real choices about whether and how to expand beyond this planet."

Deudney understands that substantially lowering the costs of accessing space will likely be sufficient to start unstoppable waves of human expansion into space:

> "If it is cheaper and easier to get to orbit, then all space activities become cheaper and easier to accomplish, whether or not their effects are desirable. The main reason so many space expansionist projects with ultimately undesirable outcomes have not yet been undertaken is simply because they

cost too much."

What does Deudney propose to do with the entrepreneurs and space engineers who are successfully striving to substantially lower the costs of accessing space? Perhaps jail?

"Coming of age means putting aside the fairy tales of childhood," says Deudney.

I think coming of age means finding ways to realize the beautiful dreams of childhood. Success, of course, is not guaranteed, and most people fail. But there's honor and glory in trying. Not even trying is, I think, premature death.

Deudney's conclusion is:

> "The pursuit of ambitious space expansion must now prudentially be judged to be deeply undesirable for humanity and the Earth for at least several centuries. As long as large-scale space ventures have been infeasible, it has been possible to view them as desirable without suffering any consequences from our illusions. But the time when we can escape doing the undesirable because it is largely infeasible will not last indefinitely. Eventually, and perhaps soon, we will no longer be protected from the consequences of our delusions by our incapacities, and we will suffer severely unless we learn to firmly say no."

Deudney proposes that armies of overzealous bureaucrats should outlaw all space projects besides current practical applications, Earth and climate science, and some space science. But, he says, it "is vital to conduct such scientific ventures in a way that prevents exploration from sliding into colonization." For example, scientific outposts on the Moon should be OK, but with robots instead of people. However, Deudney seems willing to consider exceptions and maybe allow a few humans on the Moon. Maybe.

Deudney emphasizes that space expansion could cause wars and unleash weapons of mass destruction. He is right of course. He mentions ideas like stockpiling last resort nuclear weapons in deep space to avoid detection, deploying directed energy weapons (such as military lasers) in space, or redirecting asteroids to hit enemy nations on the Earth with "planetoid bombs."

But I think there will be more and worse wars if we remain confined on this planet, with increasing population and decreasing resources.

Deudney says that, contrary to the dreams of libertarian space enthusiasts, future space communities are likely to be heavily regulated and authoritarian, at least initially. I guess he is probably right here, but I think the goal justifies the risk and I hope our space communities would gradually become less authoritarian with time.

Deudney's vision for our future on this (and only this) planet is a permanent lockdown in a retirement home. We should embrace the lockdown, drink tea, and pretend to be happy.

But recent events have shown what happens to people locked down in a retirement home: many become sick and die in pain and loneliness.

I firmly say no to Deudney's planetary lockdown.

I'm not necessarily against waiting some more years to return to the Moon sustainably and permanently, and then move on to Mars and beyond.

In *"Star Maker"* [Stapledon 1937], Olaf Stapledon tells the story of "a race which was not a single species but an intimate symbiotic partnership of two very alien creatures." Stapledon's aliens will eventually play a very important role in the future of the galaxy. But the two symbiotic species go through an extinction-level war against each other before expanding beyond their home planet.

The crisis is averted at the last moment, and "the instruments of a lean but hopeful civilization were refashioned. This was a temporary civilization... but one which promised itself great adventures in the 'upper world'" as soon as it had established the basic principles of a sustainable civilization.

I'm persuaded that we must, at the same time, expand into the black sky and make the green Earth a better place for everyone.

But perhaps our conflicts are too severe for that. Perhaps - I don't think so, but Deudney and others do - we must solve our problems down here before moving up there.

But if so we must promise ourselves great adventures in space as soon as we have solved our urgent problems here on Earth. I wouldn't be against Deudney's Earth-centered space program if it included, say, a few little science bases on the Moon and Mars, permanently crewed. This would keep our great expectations alive while we establish a sustainable civilization.

Humanity "may need to preserve the dream of spaceflight," notes William Sims Bainbridge [Bainbridge 2015], "to compensate us for some of the suffering we experience here on Earth, and as a possible long-term goal."

We need a frontier beyond the horizon, where the restless can hope to go, for our collective mental health. Many of our ancestors, including many of the best and brightest, went from Europe to America in search of something new and better. Once in America, they "went West" on the ever receding frontier until the Pacific coast, as chronicled by Frederick Jackson Turner in *"The Significance of the Frontier in American History"* (1893) [Turner 2011].

It's no wonder, then, that the Pacific coast has been (and hopefully it will continue to be) fertile soil for innovation. Now there are no Earthly frontiers, and therefore we must open the space frontier. The mental benefits of the space frontier up there will likely trickle down, eventually, also to those who choose to stay home. Isn't it wonderful and energizing to think that you could go yourself one day, or your children?

I'm writing this in the spring of 2021, and millions of people all over the planet are still in pandemic lockdown. Some complain, but many endure the lockdown because (and only because) they hope to be free again some day soon.

Waiting "at least several centuries" in a planetary lockdown without hoping for a better future, like Deudney suggests, is far too much. Also, policy prescriptions for several centuries are impossible to enforce and likely to backfire. Our descendants will make their own choices, based in part on what we leave them.

In Deudney's long-term planetary lockdown, humanity would either become irreversibly senile or destroy itself like enraged rats in a small cage.

When many rats are locked down in a cage, a critical mass is reached "whereby the rats in the cage would either fight relentlessly or demonstrate some form of capitulation behavior," notes Marcel Danesi [Danesi 2018]. In other words, some caged rats kill and others lose the will to fight and live.

Decades ago John Calhoun at the National Institute of Mental Health studied caged rats [Calhoun 1962].

Some aggressive, dominant males exhibited "signs of pathology, going berserk, attacking females, juveniles and the less active males," reported Calhoun. Other males "were completely passive and moved through the community like somnambulists. They ignored all the other rats of both sexes, and all the other rats ignored them... their social disorientation was nearly complete."

Others were "hyperactive... both hypersexual and homosexual, and in time many of them became cannibalistic." Many females "were unable to carry pregnancy to full term or to survive delivery of their litters if they did. An even greater number, after successfully giving birth, fell short in their maternal functions."

Robert Sapolsky notes [Sapolsky 2017] that Calhoun's paper was hugely influential but perhaps too "colorful" and not entirely accurate. According to Sapolsky, high-density living doesn't make rats more aggressive. "Instead it makes aggressive rats more aggressive... In contrast, crowding makes unaggressive individuals more timid. In other words, it exaggerates preexisting social tendencies." This is just as bad, if you ask me.

Parallels with current events and trends are easy to spot. In fact, 2020 will also be remembered for escalating social tensions, notably in the US. Unfortunately, this trend can only be expected to continue and become even worse if things remain as they are. It could likely result in widespread random violence, small wars and big wars, worse than Deudney's space wars.

To avoid this, we must continue advancing on the space frontier toward the stars, calmly but strenuously and steadily, with great expectations of wonderful futures. I am persuaded that only the possibility to move away, even if remote, can overcome the instinct to hate others when things turn bad. Those who spend their time and energy trying to move to the frontier, or create the possibility of doing so, are less likely to hate and kill others.

People have, and will always have, different temperaments and outlooks. Some people want to change things radically, and some people want things to stay as they are. This is good if you ask me, but in a closed world radical changes bring social tensions and conflicts that can only become worse with even more radical changes.

I think everyone should have the right to make their own choices, but nobody should have the right to choose for others. In the future, there will be people who choose the Earth, and people who choose the stars. I hope we will find ways to work together to create wonderful futures on the Earth and among the stars.

Another book published in 2020, *"The Decadent Society"* [Douthat

2020] by Ross Douthat, provides a much needed counterpoint to Deudney's lockdown proposal.

In a perceptive review [Thiel 2020], Peter Thiel noted that the book "sets the stakes for the most urgent public debate of the 2020s: How do we get back to the future?"

Douthat's book begins with the closing of the space frontier and the resulting decadence of the West (perhaps to be followed by global decadence) and ends with possible pathways out of decadence. In between, politics, media, culture wars, geopolitical tensions and promises, and decadence scenarios ranging from "sustainable" to catastrophic. While both hold moderate and pragmatic political views, Deudney leans liberal and Douthat leans conservative.

Douthat is clearly unhappy with the decadence of our society, but presents other viewpoints fairly. For example, a "sustainable decadence" scenario essentially similar to Deudney's planetary lockdown, in which the whole planet becomes something like a relatively happy retirement home, could be (according to some) better than more adventurous but also more dangerous alternatives.

The last two chapters of Douthat's book explore possible ways out of decadence, sustainable or not. One is the space frontier. Douthat says:

> "I suspect that a truly globalized civilization cannot help tending toward decadence so long as it remains earthbound, so long as there is no hope of finding actual new worlds to leap toward, conquer, or explore... So if we want to really escape decadence... we need to find a way to climb, to make a ladder to the stars, and to offer future generations of humanity a new reality to explore..."

Douthat's book ends with:

> "So down on your knees - and start working on that warp drive."

Thiel criticizes this tongue-in-cheek reference to the very speculative idea of reaching the stars with faster than light warp drives [Chapter 8]. We "would do well to expect our salvation to be worked out in the solar system we have been given," he says.
I don't disagree, but I think Douthat is also referring to more conservative advances. Freeman Dyson said [Dyson 1979] that the costs of space operations must be very significantly reduced "before the large-scale expansion of mankind into the solar system will be possible," and this cost reduction process is ongoing.

Once we reach a critical cost threshold, which could happen soon, the solar system frontier will be open.

Even warp drives and interstellar spaceflight might belong to the realm of the possible [Chapter 8]. "One can be skeptical of this utopianism and still see it as a possible seedbed" for spectacular advances on this and other technological frontiers, notes Douthat.

Perhaps our enthusiastic, highly imaginative space propulsion researchers won't build a warp drive anytime soon, but still develop viable enabling technologies for the solar system frontier. And then, perhaps, why not, even warp drives for the interstellar frontier.

Douthat is persuaded that some kind of religious renaissance is needed to break free of our mental lockdown and our physical lockdown on this planet [Chapter 6]. I agree, and here's to warp drives!

4 - SPACEFLIGHT NOW

In this chapter I'll focus on human spaceflight now and for the next few decades.

Today, three important factors make spaceflight different from the Apollo years: the cost of accessing space is rapidly decreasing due to new launch systems including reusable rockets; the private sector is playing a more and more important role; new geopolitical actors, first and foremost China, are playing an increasingly more important role.

The first two factors are strongly related. In fact, today's cheaper launch systems have been developed by the private sector without prior government funding. State actors will likely catch up, but at this moment private companies like Elon Musk's SpaceX and Jeff Bezos' Blue Origin [Davenport 2018], or Richard Branson's Virgin Galactic [Schmidle 2021], have the initiative and the early mover advantage.

Cheaper access to space will enable profitable space business models beyond geostationary Earth orbit, and therefore sustainable human expansion into outer space led by private industry. Once a threshold of plausible profitability is achieved and there are solid business cases for making money by living or working in space, there's no falling back down here.

Cheaper access to space will also enable nations to pursue geopolitical and military advantages through space programs beyond geostationary Earth orbit, and therefore sustainable human expansion into outer space led by governments.

"We should be looking for technologies that will radically transform the economics of going into space," said Freeman Dyson [Dyson 1979]. "We need to reduce the costs of space operations, not just by factors of five or ten but by factors of a hundred or a thousand, before the large-scale expansion of mankind into the solar system will be possible."

Reusable launch systems have already reduced the costs of space operations by a factor or ten or so. Further improvements in technology will certainly continue to reduce the costs of space operations. Economy of scale, perhaps enabled by new markets for rapid long-distance rocket transportation of cargo and passengers around the Earth [Zubrin 2019], will likely reduce the costs even more. The reduction by a factor of a hundred envisioned by Dyson doesn't seem that far.

Besides governments and corporations. I hope citizen organizations and interest groups will play a direct role in space exploration and utilization. In [Prisco 2009] I proposed ways to revitalize global enthusiasm for space and let everyone join in. For example, I envisioned space makers all over the planet cooperating online, swapping designs for 3D-printed space hardware, and sharing test results.

I dream that global groups of space enthusiasts will be able to directly fund and execute ambitious space missions. I co-founded a decentralized autonomous organization of space professionals and enthusiasts to realize this dream [Prisco 2017].

In this chapter I'm focusing on the next few decades and keeping close to current technology. So I'm not considering any revolutionary breakthroughs in space propulsion [Chapter 7, 8].

Nuclear space propulsion is an interesting intermediate case of current technology that hasn't been used so far. Proposals to leverage nuclear technology for spaceflight cost reductions have been around for decades [Dyson 1979, Landis 2013]. In the past, however, these proposals have always been shelved due to environmental and political concerns.

But nuclear propulsion "is the only choice of mature and developed technology that can offer high specific impulse and high thrust levels simultaneously today," notes Martin Tajmar [Tajmar 2003].

> "Renewed interest in manned Mars missions may revive efforts in nuclear propulsion despite present environmental and political considerations."

Today, nuclear space propulsion is on the table again [Goswami 2020], notably in the US and China.

Besides corporate and national interests, there are other practical reasons to move into outer space. One is to protect us from existential risks: a big asteroid could hit the Earth anytime and cause an extinction-level catastrophe that we could be powerless to prevent. Or we could cause an extinction-level catastrophe ourselves. Human communities in space, large enough to ensure the survival of humanity, would be a much needed insurance policy.

Another practical reason to move into outer space is that offshoring heavy, polluting industries to outer space could be the only way to protect the environment of our planet as our energy needs continue to grow [Chapter 5].

I'm all for space tourism and I hope the money spent by wealthy space tourists, and their public appeal, will help the space industry, but I don't think space tourism will play too important a role. I could be wrong, of course.

There are many ongoing spaceflight discussions and heated debates, e.g. national or global, civilian or military, government or private sector, Moon or Mars, humans or robots.

Some people think that there's no need for humans in space, because robots are good enough. But this is wrong, because robots aren't good enough; otherwise we would have robotic firefighters instead of human firefighters here on Earth.

Of course, robots can replace humans for some routine tasks (e.g. building infrastructure in space), and play critical support roles for other tasks. But people are still very much needed. Also, robots in space don't stimulate public enthusiasm for spaceflight. People in space do.

The Apollo program in the 1960s ignited public enthusiasm for spaceflight beyond a critical mass. This hasn't been the case of later space programs such as robotic missions, the space shuttle, and the International Space Station. This shows that ambitious human spaceflight is needed to generate widespread enthusiasm and public support for space programs.

Some space enthusiasts are persuaded that we should skip the Moon and go straight to Mars. I totally disagree. Focusing space programs on distant (in space, time, and money) goals can only ensure that space policy remains empty talk with no action. Focusing instead on useful products attracts money and talent that will build the infrastructure to eventually achieve distant goals.

To me, returning to the Moon to stay, building a sustainable human presence on the Moon, and bootstrapping lunar industries, are the obvious next steps in human spaceflight. Our first priority must be "to kick off the industrialization of the geolunar space," argues Adriano Autino [Autino 2020], co-founder and president of Space Renaissance, a global citizen organization of which I'm a member. "Everything else can be discussed later." I also joined the Moon Village Association, a global citizen organization of space professionals and enthusiasts focused on this goal.

Mars can wait a bit. Going back to the Moon is the wise thing to do at this moment. While there's no solid use case for Mars today, promising use and business cases for returning to the Moon and utilizing lunar resources, with reasonable costs and reasonable expectations of return, are beginning to appear [Spudis 2016].

But space companies aren't yet rushing to the Moon with their own money, without government support. There is, in fact, a big difference between the "promising business cases are beginning to appear" and "solid business cases exist" phases. In the first phase, a few venture capitalists start investing some little money, but most large investors stay behind to see what happens. Lunar based industries are still in the first phase, and still need public funding to bootstrap.

At this moment, many space companies seem mostly interested in being government contractors. This shows that governments must continue to fund ambitious space initiatives until the time is right for industry to step in, which is not quite yet.

The space frontier is often compared to the American West frontier [Turner 2011]. While American settlers "were doing so as free individuals, not as agents of a centrally controlled government" [Bainbridge 2015], their frontier was opened by the central government with suitable bootstrap initiatives including initial explorations and incentives. This supports the idea that governments have a key role to play until free agents are ready to settle the new space frontier.

Governments should lead the way, and encourage private industry to step in when the time is right. A parallel is found in the history of the internet, which was first developed with government funding. Eventually, the first prototype of the World Wide Web was released (worth noting, by researchers at CERN, a large public lab), and solid business cases materialized overnight. Then commercial players stepped in, and today's connected world is the result.

In the US, the Trump administration launched the Artemis program and instructed NASA to land the next astronauts on the Moon in 2024 [Goswami 2020]. I feared that, following the usual pattern [Chapter 2], the Biden administration would trash Artemis immediately after being sworn in. I'm pleasantly surprised that, at the time of writing (spring 2021), this hasn't happened.

Artemis is still on the table, with the target date of 2024 likely to be pushed to the end of this decade.

This would be good enough for me. I never thought 2024 was feasible in the first place, and I'm quite happy thinking that the next astronauts will walk on the Moon by the end of this decade, with Mars and the rest of the solar system to follow.

I recommend two recent science fiction novels that picture the first phase of human expansion into the solar system. "*Delta-v*" [Suarez 2019], by Daniel Suarez, is a story of early commercial spaceflight and asteroid mining. "*Ad Martem 12*" [Bassani 2018], by Giulia Bassani, is a delicious coming of age story. Bassani, a young author worth watching, tells the story of the first teenagers born on Mars.

I think that becoming an interplanetary and then interstellar species is our cosmic destiny and existential imperative. Therefore, I'm a space expansionist [Deudney 2020]: I firmly support humanity's expansion into space as a primary goal. However, I realize that many don't share my philosophical convictions, and that I can only persuade them with arguments based on practical utility and returns.

I'll now outline some promising use and business cases that could (and should), in my opinion, motivate both governmental and commercial interest in lunar resources and a following wave of human expansion into the solar system.

Lunar resources can be sent to the Earth and used cost-effectively on the Earth, or sold at a profit. The Moon can be mined [Spudis 2016, David 2019] for rare earth elements, helium-3, and water.

There is a high demand for "rare earth elements vital to defense and high-technology industries" [David 2019]. At this moment, China controls more than 80 percent of the global rare earth elements supply, which could lead to dangerous geopolitical developments and the need to pursue lunar supplies independent of

China.

Helium-3, which is very scarce on the Earth, can be mined on the Moon and shipped to the Earth to power next-generation nuclear fusion reactors. Lunar helium-3 could open the door to environmentally safe nuclear fusion, which might be the solution to future energy needs [Chapter 5].

I think the best way for the government to create cheaper space transportation options suitable for lunar industries is to support commercial operators. Instead of developing new government-owned space transportation systems, which tend to be killed by petty political squabbles and cost overruns before they could have launched much, the government should buy lunar transportation and cargo services from commercial operators.

Lunar water and the oxygen that can be extracted from it could be used by people living in lunar outposts. Furthermore, lunar water can be processed into rocket propellant for future crewed missions to Mars and other planets. Launching deep space missions from the Moon would be much cheaper than launching from the Earth, and the Moon would be a gas station on the way to the stars [Goswami 2020].

Of course, these arguments based on lunar water are somewhat circular, since they assume the need for crewed lunar outposts and crewed deep space missions in the first place.

However, a comprehensive open study of a commercial lunar propellant architecture [Kornuta 2019] demonstrates, according to its authors, "both the technical and economic feasibility of establishing a commercial lunar propellant production capability," with the assumption that the US government will purchase the propellant to "fuel a new age of economic expansion, sustained space exploration, settlement, and American leadership in space." The proposed production facility would be, at least initially, purely robotic with no human presence required.

After the Moon, and perhaps before Mars, near Earth asteroids will be a sensible next step. After Mars, the asteroid belt between Mars and Jupiter will be a sensible further step. It's worth noting that, as suggested by Gerard O'Neill [O'Neill 2013], asteroids could be converted into space habitats for large human populations [Chapter 7].

The use and business cases for mining the asteroids [Lewis 2014, Suarez 2019] are similar to those for mining the Moon. Some space enthusiasts are persuaded that we should skip the Moon and go straight to the asteroids, but I disagree. I think the Moon will be an ideal launch base for sustainable asteroid mining programs. Nuclear powered spacecraft could be launched from the Moon first, without environmental concerns.

In *"Scramble for the Skies"* [Goswami 2020], Namrata Goswami and Peter Garretson provide a detailed analysis of current geopolitical aspects of spaceflight. Besides America and China, the authors discuss India and two small nations that have introduced legislation to attract commercial spaceflight operators: Luxembourg and the United Arab Emirates.

Based on current trends and technologies, Goswami and Garretson analyze possible spaceflight futures in the inner solar system up to the asteroid belt, and build different scenarios that could play out between now and 2060:

> "The most familiar scenarios are those with a United States or China-centric world order or a bipolar U.S-China peer competition. The least familiar scenarios are those where India is the leading power or bipolar scenario where India is among the leading states... The authors' expected scenario assumes that the United States fails to initially mount a comprehensive response, allowing China a strong lead toward space resources, followed by a United States over-reaction to attempt to regain leadership..."

The authors are persuaded that India is likely to become a space superpower alongside the US and China. I first found the inclusion of India surprising. But India's population is younger and growing faster than China's. India has a vibrant high tech industry and an ambitious space program. I'm very fond of India's cultural and spiritual traditions [Prisco 2020], and therefore I think this is good. I find the exclusion of Russia surprising, and I wouldn't write Russia off just yet.

Goswami and Garretson emphasize that China could seize the initiative and achieve space supremacy over the next few decades.

For that, China is developing full stack space capabilities including a wide range of launchers for all space destinations in the inner solar system, low and geostationary Earth orbit satellites and infrastructure for civilian and military applications, crewed spaceflight, lunar and planetary landers, lunar and asteroid mining, space robotics, and deep space exploration.

Plausibly, China wants to establish cislunar space supremacy first, with first mover advantage and a de-facto monopoly in lunar industry, first and foremost mining. Mining lunar rare earth elements and helium-3, and preventing others from doing so, could establish China's de-facto monopoly on critical resources.

China is researching space solar power systems [Chapter 5] to collect solar energy in space and beam it to the Earth. Nuclear fusion powered by lunar helium-3 [Chapter 5] could become a reality in a few decades. These developments could enable China and its allies to dictate terms to a world strangled by energy needs growing faster than supply. Of course, China would exploit this position for geopolitical advantages.

Both China and the US are persuaded that the lunar poles "convey tremendous strategic advantage within the presently understood possibilities of technology."

A 2019 report quoted in [Goswami 2020] warns that China intends "to industrially dominate" cislunar space for space-based manufacturing, resource extraction, and power generation. A 2020 US Department of Defense report [Defense 2020] outlines possible military implications of Chinese supremacy in cislunar space. The creation of the US Space Force likely indicates that China's threat is taken seriously in the US.

Garretson and Goswami confirm that developing China's capacities in cislunar space is a critical component of Xi Jinping's "China dream" [Chapter 2], "and directly connected to maintain China's access and dominance in space-based assets that can be used to strengthen its earth-based military."

Bill Gertz bluntly emphasizes [Gertz 2019] that China could use space superiority for military operations in its quest for global supremacy.

I'm persuaded that, at this moment in history, only a near and credible threat from China could stimulate the US to vigorously push ahead with Artemis and other crewed space programs beyond geostationary Earth orbit. Therefore, I welcome China's push to outer space.

This book is a call for the US and the rest of the West to react. I hope that we will react. Otherwise, Xi will own the Moon and the solar system, and many space enthusiasts all over the world, including me, will welcome Xi's leadership.

5 - SPACESHIP EARTH

We spaceflight enthusiasts should see ourselves as crew members of Spaceship Earth en route toward our cosmic destiny. To advance on the road to the stars, we must strenuously protect our spaceship and its crew. Buckminster Fuller warns [Fuller 1969] that we must not only stop "mis-using, abusing, and polluting" the life support system of Spaceship Earth, but also start redistributing wealth and opportunities fairly among all crew members.

Many spaceflight enthusiasts, including me and hopefully you, totally agree. Protecting the environment of our green Spaceship Earth, and promoting social justice, are integral parts of our space mission. Conversely, spaceflight can help achieve a greener and nicer Earth.

Space cadets, tree huggers, and social justice warriors (to use three common pejorative terms) should be allies, or at least coexist peacefully, because we share the same cultural DNA. Steven Wolfe argues [Wolfe 2015] that we must take care of each other, value the life of everyone, make sure that everyone has what they need, take care of the Earth and its biosphere, and expand beyond the Earth to bring life to the stars. These are our obligations to humanity, life, the Earth, and the universe.

Legendary environmentalist James Lovelock introduced the idea that the Earth can be regarded as a whole living planet [Lovelock 2000], which he refers to as Gaia after the Greek goddess of the Earth. According to Wolfe, our duty to Gaia is to bring her seeds to the stars.

But many environmental and social activists oppose human

spaceflight. Why? And how can we persuade them that we should be allies (or at least coexist peacefully)?

I like good ideas, but I dislike ideologies. Lamentably, ideas tend to come in ideological packages that one is supposed to swallow whole and defend at any cost. So if you like A you must also like B because it is in the same package, and you must hate C because it is in another package.

Many environmental and social activists see human spaceflight as part of the other package - the enemy package. These people often mean well, but they are entirely wrong. Spaceflight is not only compatible with, but a key enabler of, the twin visions of a green Earth and a fair society.

They think most of science fiction, especially Golden Age science fiction, is part of the enemy package. But in *"Our Angry Earth"* [Asimov 2018], Isaac Asimov and Frederik Pohl, two iconic Golden Age science fiction authors, bluntly and forcefully address the need for urgent action to protect the environment of the Earth. The 2018 edition of the book, first published in 1991, has a new introduction and a new afterword by Kim Stanley Robinson, another science fiction writer especially concerned with the environment.

Besides the recognition that the space program "is, by and large, one of the most wonderful and promising endeavors ever undertaken by the human race," there's not much about space in *"Our Angry Earth."*

However, the cover of the 2018 edition of the book features a great statement by science fiction master Arthur Clarke: "A crash course in saving Planet Earth - our only home for generations, if not centuries, to come." Clarke's statement emphasizes complementary truths: one day human civilization will have to massively expand beyond the Earth, and in the meantime we must protect the Earth.

Both Greta Thunberg and Elon Musk work toward a greener Earth in their own different ways. Greta inspires young people to protect our planet. Elon builds electric vehicles and develops related technologies that contribute to protecting our planet. It seems to me that those who love one should also love the other. But no, many people who love one hate the other for "belonging to" a different ideological package.

At this moment I'm agnostic about catastrophic climate change predictions. I guess they are worth listening to, but perhaps overblown. The thing is, I'm not an expert myself, and I don't entirely trust scientists when it comes to politically overcharged issues [Crichton 2004], because I tend to assume that loyalty to one or another ideological package takes precedence over science. However, we can only persuade climate change activists to support our cause if we concede that their arguments are valid, which is what I'll assume in the following.

The green opposition to nuclear energy is another good example of the toxic effects of ideological packages. Fortunately, some reasonable greens speak up.

"I realize that for many readers, the notion of supporting nuclear technology will be a difficult one," says Ramez Naam [Naam 2013]. "Until solar and wind are paired with future storage technologies to provide 24/7 energy, nuclear is our best hope for reducing the amount of carbon we emit."

"Nuclear energy," Greenpeace co-founder Patrick Moore told Naam, "is the best technology we have today to replace fossil fuels and reduce greenhouse gases."

Legendary environmental activist Stewart Brand used to oppose nuclear energy. But then his opinion on nuclear "flipped from anti to pro," he says in *Whole Earth Discipline* [Brand 2010]. "The question I ask myself now is, What took me so long? I could have looked into the realities of nuclear power many years earlier, if I weren't so lazy."

"I came to regret leaving fusion out of my nuclear chapter," says

Brand in an afterword. "I bet that fusion will be largely welcomed by Greens, if it comes to pass. Legacy resistance against old new tech continues, but new new tech appears not to arouse the fears and activism of old."

"By all means, let us use the small input from renewables sensibly, but only one immediately available source does not cause global warming and that is nuclear energy," says Lovelock as reported in [Brand 2010]. "Opposition to nuclear energy is based on irrational fear fed by Hollywood style fiction, the Green lobbies and the media. These fears are unjustified, and nuclear energy from its start in 1952 has proved to be the safest of all energy sources..."

> "I am a Green and I entreat my friends in the movement to drop their wrong-headed objection to nuclear energy."

"My track record shows that I have always seen nuclear energy as a force more for good than for harm," says Lovelock as reported by John Gribbin [Gribbin 2009]. In fact, Lovelock has consistently and forcefully championed nuclear energy in many books. "I feel strongly that our need for energy should be treated as a practical problem of engineering and economics, not politics," he says in his last book "*Novacene*" [Lovelock 2019].

> "I feel equally strongly that the best candidate to supply these needs is nuclear fission, or, if it becomes available cheaply and practically, nuclear fusion, the process that sustains the heat of the Sun."

According to Lovelock, "we must move to using nuclear energy temporarily until we can either harvest solar energy efficiently or find out how to use the almost infinite supply of nuclear fusion energy."

Bill Gates argues for nuclear energy in his recent book "*How to Avoid a Climate Disaster*" [Gates 2021]. According to Gates, "it's hard to foresee a future where we decarbonize our power grid affordably without using more nuclear power." The book, written from the perspective of an engineer who wants to solve problems, is full of solid and clear proposals to fight climate change.

"Although it's still in the experimental phase, fusion holds a lot of promise," adds Gates.

Nuclear fusion [McCracken 2012, Parisi 2019] is very likely to offer a clean solution to our long-term energy needs, and there's an interesting link between nuclear fusion and spaceflight.

In *"Return to the Moon"* [Schmitt 2006], scientist and Apollo 17 crew member Harrison Schmitt developed a comprehensive end-to-end plan for mining helium-3 on the Moon and shipping it to the Earth to power next-generation nuclear fusion reactors.

The fusion reaction between deuterium and helium-3, which produces a proton and helium-4, doesn't produce energetic neutrons that would damage the fusion reactor and make it radioactive. This is the aneutronic nuclear reaction that requires the least input energy, and therefore is the easiest to achieve.

A large supply of helium-3, which is only present in trace amounts on the Earth but is much more abundant on the Moon, could open the door to environmentally safe nuclear fusion. This might be the solution to future energy needs and create a huge market for lunar helium-3.

Schmitt thinks lunar helium-3 mining is a realistic business model that could be pursued by commercial companies with no or limited government support. According to his analysis, the cost of placing large payloads on the Moon must be reduced to $3.000 per kilogram (a 20-fold reduction of the Apollo cost) to make lunar helium-3 mining commercially viable. Today's reusable launch systems are making significant and steady progress toward meeting Schmitt's target cost, which now seems achievable.

In his foreword to Schmitt's book, Neil Armstrong noted that the viability of the plan depends on three substantial and important developments: a commercial fusion reactor, an efficient mining operation on the Moon, and a reliable cargo transportation system. Schmitt's plan covers all three, with plenty of detail.

Schmitt's plan has been criticized, mainly because the first of these elements, a commercial fusion reactor, doesn't exist yet. However, there are promising indications that the ITER project [McCracken 2012], and/or next-generation fusion projects [Parisi 2019], could soon achieve commercially viable nuclear fusion.

Besides Schmitt, many experts are persuaded that helium-3 fusion is the best option for our long-term energy needs. I find their arguments persuasive.

The strongest counter-argument, I think, is that future fusion reactors could use other aneutronic fusion reactions that, while requiring higher input energy and therefore being harder to achieve, wouldn't need to use helium-3. But this sounds like trying to solve the actual problems of today with the hypothetical resources of tomorrow. I think that, as soon as helium-3 fusion is proven operationally viable, mining helium-3 on the Moon will be recognized as an imperative.

Schmitt briefly covers lunar surface solar power systems as well. David Criswell proposed [Criswell 2002] to build large solar collectors on the lunar surface, using local lunar materials, and beam power to the Earth via microwaves. This is one of several space based solar power concepts.

Space based solar power systems [O'Neill 2013, Mankins 2014] would collect solar energy in space and beam power to the Earth. The potential of space based solar power systems to deliver green power to meet the world's energy needs is firmly established, and implementation plans are on the table [Goswami 2020].

Besides delivering green energy, space based solar power systems could be used to directly modify the climate, for example by using giant mirrors in space to focus solar radiation onto specific regions of the Earth [Liu 2013a]. This is an example of geoengineering [Brand 2010]. According to Bill Gates [Gates 2021], "geoengineering is the only known way that we could hope to lower the earth's temperature within years or even decades without crippling the economy." Future geoengineering technologies on steroids could permit terraforming Mars and other planets [Chapter 7].

Brand notes [Brand 2010] that the environmental movement was inspired and bootstrapped by the powerful "overview effect" [White 2021] of seeing the Earth from space. Today, Earth observation satellites continuously monitor the health of our planet. Tomorrow, spaceflight will directly help achieve the goals of the environmental movement.

Spaceflight will "open for our use new sources of energy and materials while preserving our environment," noted Gerard O'Neill in 1976 [O'Neill 2013]. "We need to move heavy industry off the Earth," said Jeff Bezos in 2019, presenting the Blue Moon lunar lander and cargo system to the press. "It will be way better done in space anyway. And Earth will be rezoned residential and light industry."

Offshoring energy production and heavy industry to outer space is, I think, the only viable way to protect the environment of our planet as our energy needs continue to grow, and the only viable option for the future of humanity on this planet and beyond.

The planet could also be saved by dramatically reducing our energy consumption, or "even better" eliminating human technology altogether. Theodore Kaczynski made a famous proposal in this direction [Kaczynski 2010]. But while the Unabomber's proposal is green, it's most certainly not nice: it would cause widespread suffering and countless deaths.

One often hears that the human population is exceeding the carrying capacity of the Earth. But I'm persuaded that having children is a human right, and having more people is good [Autino 2020]. Expanding into space is, I think, the only way out of this dilemma.

I (and I hope you) aspire to a nice world where all people have a decent life and are able to pursue happiness in their own personal ways, and meaningfully contribute to the common project of making Spaceship Earth a better place.

Unfortunately, some people who also aspire to a better and nicer world are opposed to spaceflight because, here again, they see spaceflight as part of the enemy package. But it isn't.

"People whose concerns are very much focused on their own personal condition," notes William Sims Bainbridge [Bainbridge 2015], "will be far less interested in grand projects that are irrelevant to their own immediate needs." On the contrary, affluent people tend to support spaceflight more than the average population. Similarly, Bainbridge notes, educated people tend to support spaceflight more than the average population.

Therefore, all spaceflight enthusiasts should be in favor of equitable redistribution of wealth, social fairness programs, and affordable education for all. Of course this would take public funds away from space programs, but I think the resulting increase in public support would compensate in the longer term.

I have always been in favor of equal opportunity, but I have reservations about affirmative action. I feel that, since affirmative action gives some groups of people more opportunities than others, it directly contradicts equal opportunity. You can be for one, or for the other, or against both, but not for both.

A strong counter-argument is that, since opportunities have not been equal in the past and have not yet been fully equalized today, affirmative action is still needed to build a world where opportunities are really equal for everyone, in practice as in theory, regardless of their skin color, gender, sexual orientation and all that. Then, and only then, we'll be able to abandon affirmative action.

I understand this argument, but I still find affirmative action policies annoying, especially when it comes to spaceflight.

However, we live in the real world where facts are facts. And the fact is that there are some spaceflight enthusiasts, but many more social justice activists. Our beloved rockets won't take off without their support.

So I have decided to make peace with the fact that affirmative action is here to stay, up there in space like down here on Earth, and to see the bright side of this.

Making humanity interplanetary and then interstellar will require the effort of a lot of people, the more the better, over many, many generations. We really need everyone's contribution.

Young women and young people from disadvantaged minorities don't have, yet, enough role models to make them dream of becoming space engineers or astronauts and put in many years of hard study and work to achieve their dreams. Seeing others like them lead NASA, or walk on the Moon, could make all the difference and unleash human potential that is still hidden.

The NASA videos of today, with people of all genders and colors, inspire more people than previous videos and win more support for spaceflight.

As far as I'm concerned, NASA should feel totally free to send only female and minority astronauts to space. I'll avidly follow and promote them on social media, and so should you.

So if you are one of those space enthusiasts who, like me, tend to find affirmative action and too much "woke" identity politics annoying, my message to you is: stay focused on our goal and don't complain about political correctness.

Don't complain if women and minorities are given priority for high profile space jobs in the public sector. Don't complain if they get to walk on the Moon and Mars before other equally qualified astronauts. If you are a science fiction fan, don't complain if they win all the Hugo awards. Think of this as the price that you must pay (there ain't no such thing as a free lunch) for public and political support for spaceflight.

Conversely, the social justice activists who oppose human spaceflight should realize that astronauts and space engineers are and will continue to be powerful symbols of human excellence. Those who really want social justice should realize that, alongside other rights, the right to dream of becoming a protagonist of history is an important human right.

6 - CULTURAL REFLECTIONS

I watched "*2001: A Space Odyssey*" [Benson 2018, Clarke 1968], by Stanley Kubrick and Arthur Clarke, a few weeks before Apollo 8 orbited the Moon and a few months before Apollo 11 landed on the Moon. The film has inspired my generation, and younger generations, with the cosmic sense of wonder of the best science fiction and a burning enthusiasm for spaceflight.

For a few years the Apollo program encouraged hopes that, by 2001, we would have large bases on the Moon and human spaceflight to Jupiter and Saturn. But we didn't get the 2001 of Clarke and Kubrick. We got September 11, 2001 instead, which set the tone of Western culture to pessimism and defeatism.

Now it seems that perhaps, maybe, human spaceflight to the Moon and beyond could restart. But there will be impediments and interruptions on the road, and too optimistic expectations will be frustrated again and again. We need, instead, a vibrant spaceflight culture sustainable for the long run, and a futurist spaceflight philosophy.

In this chapter I'll discuss the intersections and interactions of spaceflight, popular culture, and philosophy.

"*Of a Fire on the Moon*" [Mailer 1970], the chronicle of the Apollo 11 adventure written by Norman Mailer, is a literary masterpiece that captures the larger-than-life spirit of Apollo.

Of course, Mailer also noticed the cracks in the Apollo narrative, some problematic aspects of which were condemned by poet and singer Gil Scott-Heron in "Whitey on the moon" [Scott-Heron 2000].

Mailer's pictures of white working class Americans going en masse to Florida to proudly watch the Apollo 11 launch are vivid and unforgettable.

But Whitey is on the Moon no more, and it seems likely that one reason is that Whitey went alone without the rest of the world. Space doesn't belong to Whitey: it belongs to everyone, and there must be narratives for everyone.

Early space narratives were centered on the "right stuff" [Wolfe 1979] owned by top fighter pilots and test pilots, from whose ranks the first astronauts were chosen. The right stuff is related to physical strength and endurance, toughness, courage, determination, ability to perform under extreme pressure, and similar things.

Not everyone has the right stuff (for example, I don't). This is a problem, because public support for spaceflight can only be sustainable if there's something for everyone. This doesn't make the right stuff wrong, on the contrary the right stuff is still very much right and needed, but it means that other equally right stuff must also have a place.

We need new narratives for spaceflight - not to replace older narratives, but to make the overall narrative inclusive. Works like *"Hidden Figures"* [Shetterly 2016] and NASA's decision to rename its headquarters building after Mary W. Jackson, the first African American female engineer at NASA featured in the novel and the derived film, are steps in the right direction.

According to a common but very misleading perception, spaceflight is something for white men.

But another literary masterpiece on the Apollo adventure, first published in the mid sixties before Mailer's book, was written by a woman.

"*If the Sun Dies*" (1965), by Oriana Fallaci, "is certainly one of the strangest, most literate, most honest, and, ultimately, most moving" books about spaceflight, according to a *New Yorker* review [Healey 2019] that praises "the almost fictional quality of the book's format and style" alongside its accuracy and frankness.

Fallaci, a journalist and often controversial opinion writer deeply immersed in the here and now, was initially skeptical of spaceflight but became an enthusiast in the course of writing the book.

A 2014 Italian edition of the book [Fallaci 2014] includes excerpts from an interview with Fallaci, who said (my translation) that the book "is a love story."

> "Perhaps the Sun will die, perhaps we'll have to migrate to other planets, perhaps we'll change, we won't be human with the body we have, the shape we have, but a shape we can't even imagine... but like it or not, want it or not, that future is coming. The Moon, other planets and other solar systems, we want to go there, and we will go there... This means preparing to pack up and go to other planets warmed by other suns: to live there when our Sun will die and the Earth will die. It means to continue Life. It means to love Life."

Another simple and clear formulations of the spaceflight philosophy that informs this book is due to a black woman:

> "The destiny of Earthseed is to take root among the stars."

Earthseed is the fictional philosophical and religious movement described by science fiction master Octavia Butler in her novels "*Parable of the Sower*" and "*Parable of the Talents*" [Butler 1993, 1998].

Butler's heroine, Lauren Olamina, is a strong willed, larger than life black woman who wants to spread the seeds of humanity among the stars.

After extreme hardship in a very dystopian near-future America,

Lauren wins:

> "She saw the first shuttles leave for the first starship assembled partly on the Moon and partly in orbit."

The story was meant to continue in other books, but was interrupted by Butler's untimely death.

Butler's biographer Gerry Canavan inspected Butler's writing notes for the planned sequels [Canavan 2016]. The adventures of Lauren's followers among the stars would be full of struggles and setbacks. Butler was no utopian, and she had mixed feelings about mostly everything including spaceflight. But space "could be our future," Butler says through Lauren [Butler 1993].

> "I believe that. As far as I'm concerned, space exploration and colonization are among the few things left over from the last century that can help us more than they hurt us. It's hard to get anyone to see that, though, when there's so much suffering going on just outside our walls."

Earthseed has inspired Martine Rothblatt, a wealthy space and biotechnology entrepreneur, an outspoken social change activist, and a transgender, to create a spaceflight oriented philosophical movement called Terasem ([Chapter 9], [Terasem 2012]).

So those who keep saying that spaceflight is something for white men don't know what they are talking about. Spaceflight is for everyone.

We need powerful space narratives to re-ignite public enthusiasm and support for space exploration, and human expansion into space.

We need cultural enablers, a new space culture, and a popular philosophy of spaceflight, to keep our energy and motivation as we walk the first long miles of our slow hike to the stars. We should create "a culture centered on human expansion into the solar system, and onward to the stars" [Benford 2013].

The internet, which has been developed in the last few decades "instead of" (so to speak) human spaceflight, is a key cultural enabler. The internet revolution has given spaceflight enthusiasts all over the planet unprecedented opportunities to connect and work with other spaceflight enthusiasts, regardless of geographic location. Once a critical mass is achieved, great things will happen.

Among other demographic factors that affect public support for spaceflight, William Sims Bainbridge notes [Bainbridge 2015] that women tend to be less supportive of human expansion into outer space than men.

I'm very skeptical of gender based generalizations, but I see that many (not all) women have little patience for abstract ideas divorced from the practical reality of everyday life. In the late 1980s and early 1990s I was an early internet evangelist, and I wondered why so many women didn't want to listen to me when I talked of the wonderful connected world of tomorrow (today).

But things changed rapidly and radically when the internet achieved a critical mass and became of practical use. Then, many women became enthusiastic internet users overnight. Today, many women are enthusiastic about the "unexpected space age" [Cruddas 2020] of location based services enabled by satellites in space. Similarly, I think many more women will support human expansion into outer space when the first practical benefits will materialize.

Science fiction has been the most powerful way to establish powerful space narratives, and a key cultural enabler. White Western men have produced great science fiction works, and I love their science fiction, but we need more voices. Today, many women, writers from minority groups, and writers from other regions of the world, including developing nations, are writing great science fiction.

China is an ascending superpower, and therefore I guess Chinese science fiction doesn't qualify as science fiction from developing nations. However, China is likely to play a leading and perhaps dominant role in spaceflight over the rest of this century [Chapter 4], and Chinese science fiction is a powerful cultural enabler for China's and humanity's future in space. Therefore, Chinese science fiction deserves special attention.

I'm a great fan of Chinese science fiction master Liu Cixin [Liu 2013, 2016a], whose *"Three Body Problem"* trilogy [Liu 2014, 2015a, 2016b] has made Chinese science fiction globally popular. A personal favorite of mine is his short story "Sun of China" [Liu 2013a].

I have been reading many other Chinese science fiction writers. In translation, because I don't understand Chinese (if I were younger, I would definitely want to learn Chinese). For a first introduction to Chinese science fiction, see two collections of translated short stories and essays [Liu 2016, 2019] and look for other works by the authors represented.

Optimism is fading out in today's West, but is alive and well in China. "The China of the present is a bit like America during science fiction's Golden Age, when science and technology filled the future with wonder," says Liu Cixin [Liu 2016a]. The optimistic sense of wonder of Golden Age science fiction [Nevala-Lee 2018] inspired the scientists and engineers who developed today's technology in the West, and will continue to inspire the developers of tomorrow's technologies.

Unfortunately, in today's Western culture, pessimism and despair are cool and optimism is uncool. But listen to Guillermo del Toro, who affirms that optimism is "the hard choice, the brave choice" [Del Toro 2019].

> "Optimism is not uncool; it is rebellious and daring and vital."

I hope Western science fiction, which is too often pessimistic, defeatist, and uninspiring, will recover its original daring optimism and inspire a spaceflight renaissance in the West. Let's foster an optimistic, daring, and vital popular philosophy of spaceflight. It will enable us to enjoy our little steps on the long road to the stars while strenuously working toward the next steps and the overall long term vision.

A couple of years ago *Vice* asked a group of interdisciplinary thinkers about their fears and hopes about the future [Ferreira 2018]. Here's my response:

Fear: the trend, evident in the "Western world," toward a senile society of sedated, reasonable, boring, politically correct zombies. Hope: humanity has done wonderful things on Earth and can move on to do even more wonderful things among the stars, provided we keep a healthy reserve of boundless, irreverent, and unreasonable optimism.

In the 1960s, American kids wanted to be astronauts (or at least scientists and engineers), and kids all over the world looked at America as the promised land of our future in space. I was one of those kids in Europe. Now I realize that, for the foreseeable future, Europe can only be a minor player in spaceflight programs led by other players. I would prefer to follow the US, but if the choice is between space expansion led by China and no space expansion, I'll be happy to follow China.

In *"Destined for the Stars"* [Newell 2019], Catherine Newell traces the roots of the Apollo space program in the popular culture of the 1940s and 1950s, which was stimulated by the works of visionary cultural engineers like Chesley Bonestell, Willy Ley, and Walt Disney, supported by equally visionary space engineers like Wernher von Braun.

Newell offers an actionable insight: the Apollo program faded out because, at the end of the day, "this version of the space program wasn't the future Americans signed up for."

> "In America's popular imagination... space was populated by them, by regular people who within a generation or less were going to transplant their suburban lifestyles to the Moon, to Mars, to Saturn's moon Titan... They were supposed to be living in one of Bonestell's pictures - standing on Mimas looking at Saturn, setting up a research base on the Moon, driving their vehicles around Mars, living in that beautiful white halo of a space station..."

The core difference between Apollo and the future in space that Americans wanted is, according to Newell, that Apollo wasn't paving the way to suburbs on the Moon and farms on Mars anytime soon.

Perhaps space enthusiasm in America faded out also because, Newell says:

> "Americans discovered that there were many other ways to work toward a future utopia besides space exploration, such as civil rights, women's rights, or the environmental consciousness..."

But space exploration doesn't have to be in conflict with those other worthy goals [Chapter 5]. On the contrary, it can support them.

Newell emphasizes that the cultural and space engineers who imagined our future as space settlers were moved by "a deep-seated belief in a sense of divine destiny to reach the heavens" analogous to the 19th century belief in the God-given "manifest destiny" to settle the American West. She notes that, according to von Braun, "it was humankind's sacred duty to leave Earth and colonize other planets."

Similarly, Roger Launius argues [Launius 2019] that human spaceflight may be viewed as a religion. The sense of wonder of science fiction "is akin to religious awe," notes Bainbridge [Bainbridge 2015], "and in it people may find at least a taste of the spiritual uplifting and cosmic meaningfulness once gained in church."

Sustainable human expansion into space requires the long term thinking "that built cathedrals in Chartres and Salisbury and Cologne," argues Paul Gilster in *"Centauri Dreams"* [Gilster 2004]. "Just because it takes centuries doesn't mean we aren't making progress," says Annalee Newitz [Newitz 2013].

> "We're riding a slow, powerful wave that will bear future generations to the stars."

I am persuaded that becoming an interplanetary and then interstellar civilization is our cosmic destiny ([Chapter 9], [Prisco 2020]). But the way to the stars will be long and full of roadblocks and setbacks. Therefore we'll need to cultivate the mindset of the cathedral builders, who worked strenuously toward awesome ends they wouldn't see.

To start moving toward our cosmic destiny, I think we should follow Newell's suggestions, go back to the roots, and promote space as a new frontier, open not only to Americans but to everyone on the planet. In parallel, we should promote the spiritual and artistic aspects of the space frontier.

Spaceflight pioneer Konstantin Tsiolkovsky was part of a philosophical and spiritual movement known as Russian cosmism [Young 2012, Groys 2018, Prisco 2020] that bloomed in late 19th century and early 20th century Russia. Later, Russian cosmism informed and energized the Soviet space program.

Russian cosmism is related to Russian futurism, an artistic movement inspired by the Italian futurist movement of Filippo Tommaso Marinetti [Rainey 2009, Campa 2012]. Contrary to some allegations, Italian futurism and fascism were essentially incompatible, as emphasized by top representatives of both [Campa 2012]. The Italian futurists were artists with little patience for politics. They celebrated vital energy, novelty, and speed. Of course, they were enthusiastic fans of aviation and early ideas of spaceflight.

Today, they would be enthusiastic fans of spaceflight.

While I don't approve of some exuberant excesses of the Italian futurists, I think today's sedate Western culture could use an injection of vitality, and I stand with Marinetti on the summit of the world to "fling our challenge to the stars" [Rainey 2009]. See [Chapter 9] for more on Russian cosmism and Italian futurism.

As previously mentioned [Chapter 3], in *"The Decadent Society"* [Douthat 2020] Ross Douthat explores possible ways out of decadence. One is the space frontier.

Another is a religious renaissance. It could come from "a general belief in the immanent divine... that's interwoven with the material world rather than standing outside it as a Creator God." But Douthat, a Christian, thinks that a religious renaissance is more likely to happen inside a traditional religion like Christianity. A powerful Christian renaissance could be stimulated by current developments in the geographic or philosophical fringes of Christianity. Douthat suggests that African Christianity could play an important role.

A space renaissance and a religious renaissance could be parallel pathways out of decadence, but also convergent pathways (as suggested by, among others, Bainbridge, Launius, and Newell). Douthat notes that there can be a "mysterious alchemy" between science and religion, and "nothing will be a surer sign that decadence has ended in something like a renaissance than if that alchemy suddenly returns."

7 - INTERPLANETARY

As noted in [Chapter 4], Namrata Goswami and Peter Garretson limit their analysis of current spaceflight and likely future scenarios [Goswami 2020] to the inner solar system up to the asteroid belt and to the next four decades, assuming only expected moderate improvements in spaceflight technology.

This is wise, because our ability to build credible scenarios would likely break down beyond these limits. Wild cards, unexpected breakthroughs, but also unexpected roadblocks, could materialize anytime and probably will materialize before the end of this century.

But of course we want to imagine what could happen after the next few decades. Fortunately, there's plenty of good science fiction that can guide our imagination.

A solar system wide human civilization has been imagined by many science fiction writers, with plenty of detail. But spaceflight has spectacularly increased our knowledge of the solar system in the last few decades, and therefore most vintage science fiction is not up to date.

Let's take a look at the interplanetary civilization scenario portrayed by James Corey (the pen name of a collective of two authors) in *"The Expanse"* [Corey 2011-2021] and the derived, very popular, TV show.

All too often film and TV adaptations of science fiction literature are dumbed down and changed beyond recognition, but the TV show stays reasonably close to the novels, at least until Season 5, with only minor changes that, I guess, are needed for TV adaptation.

"*The Expanse*" is science fiction at its best, able to provoke a sense of wonder and great expectations for our future in space. The TV show has given its many fans credible and inspiring pictures of our future in the interplanetary expanse, and I guess many fans will support current and future spaceflight programs.

The novels don't give dates, and the only date given in the show is the "since 2307" that appears on the label of a bottle of "Ganymede gin." However, it seems likely that the story begins in the mid 24th century, 300 years and a few decades in our future.

I'll first give a short outline of Corey's interplanetary civilization scenario, and then examine its plausibility.

People live on Luna (the Moon), Mars, many moons of the outer planets, artificial habitats, and asteroids like Ceres in the asteroid belt (the Belt). The main powers in the solar system are the United Nations, Mars, and the roguish Outer Planets Alliance (OPA). The OPA, which started as a labor union, controls parts of the Belt.

There's a complex interplanetary economy and commerce network, with raw materials and products shipped all over the solar system.

The Epstein Drive, named after the inventor of "the first fusion drive that solved the heat buildup and rapid fuel consumption problems of constant thrust," has opened the solar system frontier. We aren't told much about how the Epstein Drive works. In an interview included in [Corey 2011], Corey just says that the Epstein drive works "Very well. Efficiently." Corey wanted everything "to be plausible enough that it doesn't get in the way" but without too much technical detail. "This is working man's science fiction."

There's a cold war in the solar system, with occasional use of force and terrorism. I found the many action and combat scenes boring, but I guess they contribute to making the TV show and the novels appealing to many viewers and readers. Here's to action and combat scenes, if they make good science fiction popular.

I found more interesting the vivid pictures of human societies in the solar system, and the overall political scenario. Human expansion in the solar system won't result in a heaven-like utopia. On the contrary, things will be messy and often unfair (but also wonderful) just like, you know, here and now on the Earth.

The United Nations governs 30 billion people on the Earth and all over the solar system. 20 million people live on the moons of Saturn, and 45 million people live on the moons of Jupiter. There are yet unrealized plans for a network of floating cities in the atmosphere of Venus, and yet unrealized plans for space elevators for Mars and the Earth. A few thousand people live on a moon of Uranus, the farthest outpost of human civilization at the beginning of the story.

50 to 100 million people live inside hollowed homesteaded asteroids and artificial space habitats in the Belt. The Belt, "the slums of the solar system," is an impoverished and oppressed third world. While most Belters are nice people, the Belt is fertile ground for violent extremism and large scale terrorism. Eventually a Belter faction attacks the Earth with weaponized asteroids, kinetic impact weapons that cause billions of deaths.

Four billion people live on Mars. While half of the Earth's residents live "on basic" government handouts, things are very different on Mars. Bobbie Draper, a Martian, proudly thinks to herself that:

> "The entire population was engaged either directly or indirectly in the greatest engineering feat in human history: the terraforming of a planet. It gave everyone a sense of purpose, a shared vision of the future."

Those who pay attention to political correctness in science fiction should be pleased that ethnic groups and genders are equally represented, and appreciate the overall diversity and inclusivity. Jim Holden was born to a multiple marriage group with DNA from everyone in the group. The features of Naomi Nagata "showed a far-flung racial mix that was unusual even in the melting pot of the Belt." The nicest character, Anna Volovodov, is a female priest who is married to another woman.

In reply to "a tweet where someone is worried #TheExpanse is going to 'get woke'," Corey tweeted:

> "Dude, we've been woke as fuck since day one. You didn't know? Maybe you're woke too and you didn't know it."

After this short outline, let's examine the plausibility of Corey's interplanetary civilization scenario.

I think the time frame of about three centuries from now is essentially plausible. Perhaps I would have introduced more advanced Artificial Intelligence (AI) and bio/nano/neuro technology, more futuristic computing and telecom technologies, and some proof-of-concept space elevators. On the other hand, I find the idea of billions of people living on Mars a bit too optimistic in this time frame.

Many experts [e.g. Dyson 1979, Forward 1995, Tajmar 2003, Gilster 2004, Landis 2013, Parisi 2019, Zubrin 2019], are persuaded that nuclear space propulsion, and in particular nuclear fusion drives, will play a key role in our expansion into the solar system. Fusion propulsion "could be capable of taking humans to Mars in as little as 20 days, to Titan (the largest moon of Saturn) in 150 days, and our neighboring star systems in 75 years" [Parisi 2019].

Nuclear space propulsion could be followed [Forward 1995, Tajmar 2003, Gilster 2004] by antimatter propulsion, solar sails pushed by the solar wind, sails pushed by laser, microwave or particle beams, or even futuristic (some say science fictional) propulsion technologies able to use "zero-point energy" extracted from the vacuum.

The idea of creating space habitats for large human populations, either by building them from scratch or by hollowing and homesteading asteroids, has been explored by many scientists and engineers, notably by Gerard O'Neill [O'Neill 2013].

Space elevators [Forward 1995] are cable systems anchored in deep space, which would lift cargo and people from the surface of the Earth (or another planetary body) to (geo)stationary orbit very efficiently and at very low cost.

Nanotechnology [Drexler 1986, Bainbridge 2007, Prisco 2020] has produced ultra light and ultra strong materials that would permit building space elevators [Clarke 1979, 1994]. While the cost of a space elevator would be prohibitive at this moment, especially in view of the current cultural and political climate, space elevators would offer cheap and green options to access orbit, and spectacular reductions of space operation costs.

Terraforming is geoengineering on steroids that could make an entire planet like Mars habitable by unprotected humans, animals and crops [Zubrin 1997, Kaku 2018]. It's not surprising that the huge project of terraforming Mars energizes the entire Martian society in Corey's fiction.

James Lovelock co-authored with science fiction writer Michael Allaby a somewhat dated but very thoughtful fictional account of *"The Greening of Mars"* [Allaby 1984] written from the perspective of a Martian citizen in the 23th century.

"Doubtless in many of its details, Dr. Zubrin's book - like my own exercise in terraforming Mars, *The Snows of Olympus* [Clarke 1994] - will be bypassed by future advances in technology," says Arthur Clarke in his foreword to [Zubrin 1997]. It seems likely that combined advances in geoengineering, biotechnology and nanotechnology will make the prospects of terraforming Mars less daunting.

It's interesting to compare *"The Expanse"* with the interplanetary civilization scenario, set in the same time frame of about three centuries in the future, developed by Kim Stanley Robinson in

"*2312*" [Robinson 2012]. Robinson's technical, sociological, and literary treatment of future interplanetary civilization is more sophisticated than Corey's. On the other hand, I guess Corey hits the sweet spot between too clever and too dumb needed for mass appeal.

Robinson's scenario includes ultra-powerful quantum computers, strong and perhaps conscious AI, and futuristic biotechnology for human modifications and life extension. There are space elevators for the Earth (37 of them!) and Mars. Hollowed and homesteaded asteroids play the role of large spaceships for long distance hauls in the solar system.

Mercury and Venus, which are not inhabited in Corey's scenario, are also settled by humans. Mars, which has been fully terraformed in less than 150 years, is a "shirtsleeve world" with a breathable atmosphere. Venus and Titan are on the way to become fully terraformed worlds as well, and there are terraforming projects all around the solar system.

Besides more subtle methods, terraforming makes frequent use of brute force planetary engineering methods like burning the soil to release gases or using controlled ice asteroid impacts to add water. Robinson imagines, with some detail, how terraforming worked for Mars and how it is working on Venus, with vivid pictures of terraforming as it is happening.

If you find Corey too woke, you ain't seen nothing yet. Robinson will give you repeated cultural shocks with pictures of human modifications, including radical gender engineering in action, illustrated by vivid and explicit scenes. If you find these things appealing, you should welcome our interplanetary future. It seems plausible that only the possibility to move far from disliked neighbors can overcome mutual hate.

I have positive expectations for an explosion of human freedom and diversity in the solar system and then among the stars. I guess more and more people will make this cultural choice. Only, I don't want it (or other cultural choices) forced upon anyone.

Politically, Robinson's solar system is as diverse as you can expect looking at today's Earth, with many centers of power and economic systems. The Earth is politically divided among big and small nations. Of the eleven billion people on the Earth, three billion live below the poverty line and most of the others live on the brink of poverty. Mars is independent and united.

Besides the Earth and Mars, the "Mondragon Accord" plays an important political and economic role in Robinson's solar system. The Mondragon Accord is "an economic system of nested co-ops organized for mutual support" based on the (real) Mondragon system pioneered by José María Arizmendiarrieta [Azurmendi 1992] in the Basque country. Here's Robinson's vision:

> "A growing network of space settlements used Mondragon as a model for adapting beyond their scientific station origins to a larger economic system. Cooperating as if in a diffuse Mondragon, the individual space settlements, widely scattered, associated for mutual support... supercomputers and artificial intelligence made it possible to fully coordinate a non-market economy."

In [Robinson 2020], Robinson explains the real Mondragon, describing it as "an alternative to capitalism, more humane, what you might even call a Catholic political economy." At the same time, explains Tom Bell [Bell 2017], the original Mondragon Corporation is a very successful worker owned cooperative that thrives because it "aligns the incentives of the workers with those of the business as a whole."

I'm a fan of "decentralized, nonhierarchical modes of collaboration and ownership" [Eisenstein 2011] and I look forward to seeing Mondragon take roots here on the Earth, and then in the solar system and among the stars.

But more realistically, I think we can expect to see extensions of today's economy and politics in the first phase of our expansion into space. Space communities in the solar system can be expected to be established by nations and/or large traditional corporations first, and organized like the parent entity.

Daniel Deudney argues that, due to the harshness of space, future space communities will need to be highly organized and protect themselves from "aberrant behavior" [Deudney 2020]. Therefore, the first space settlements are likely to be heavily regulated and authoritarian societies. On the other hand, Charles Cockell has edited books [e.g. Cockell 2015a, 2015b] that envision libertarian space settlements.

According to Tom Bell [Bell 2017], future spacesteaders will take inspiration from the libertarian ideology of today's seasteading movement. Bell is researching the kind of legal systems humans will need once they spread across the universe.

In the long run, I think interplanetary spacesteaders of various political persuasions will pioneer all sorts of innovative forms of economic and social organization. Some experiments will fail but others will thrive, and innovation will protect human culture from ossification. Let a thousand flowers bloom in the black sky!

I have great, enthusiastic expectations for our future out there in the solar system. Of course, it will be no utopia. There'll be injustices, suffering, and wars like here and now on this planet. But I'm confident that things will be better overall. The more adventurous and restless will dream of the space frontier, and perhaps go.

Those who stay at home will look up and feel the energy of the space frontier. Back to the present, you can look at the night sky and imagine the future interplanetary civilization that we, including you and me and all other people here and now on Spaceship Earth, are preparing for. Look up [Cruddas 2020], feel the energy, and do something good.

Beyond the planets and moons of our solar system, the stars

beckon.

In *"The Expanse,"* the Mormon Church is building a huge multigeneration starship in the Belt. In *"2312,"* a small moon of Pluto is being converted to the first starships for long interstellar hauls, with engines that combine nuclear fusion and other early interstellar propulsion technologies [Chapter 8]. In both scenarios, the stars are still very far.

But Corey introduces a wild card. Extremely powerful alien nanotechnology (the "protomolecule") is found on Phoebe, a moon of Saturn, and wreaks havoc in the solar system. Eventually, the protomolecule builds a wormhole gate [Chapter 8] in the outer fringes of the solar system, through which humans will gain access to the stars.

Wild cards can happen, and at times do. We could find alien technology in the solar system, or receive a message from aliens, or even be visited by aliens. Advanced alien civilizations could have "left artifacts for man to discover" on the Moon ([Norden 1968] and of course [Clarke 1968]). Some scientists are persuaded that we have already seen alien spacecraft passing through the solar system [Loeb 2021]. But we can't count on wild cards. In [Chapter 8], I'll speculate on how our descendants will achieve interstellar spaceflight and move to the stars.

It's worth noting that large amounts of helium-3, which is efficient nuclear fusion fuel as discussed in [Chapter 5], could be mined in the atmospheres of outer solar system planets like Jupiter and Saturn [Landis 2013, Zubrin 2019]. This, and interstellar launch infrastructure in the solar system like huge lasers and beamers in space, could enable the first, relatively slow interstellar missions. Therefore, to go to the stars, "first we will industrialize the solar system" [Landis 2013].

8 - INTERSTELLAR

Some people think that we should abandon our dreams to go to the stars. Interstellar spaceflight, they say, is impossible.

I call bullshit. Don't forget that, only a few years before the Wright brothers, respectable scientists claimed that airplanes were impossible.

Advances in space propulsion will enable "even interstellar missions," notes Martin Tajmar [Tajmar 2003].

> "And there are even more fantastic ideas around that will probably enable us to master gravitation itself and revolutionize space travel…"

Incremental technology improvements and streamlined use of resources will open the interplanetary frontier. But for the interstellar frontier, we'll need better space propulsion technologies.

The book *"Frontiers of Propulsion Science"* [Millis 2009] presents recent advances in propulsion technology, and promising research steps to discover how radical propulsion breakthroughs might finally be achieved. In *"Centauri Dreams"* [Gilster 2004], Paul Gilster provides an overview that remains one of the best references on interstellar spaceflight, including the propulsion technologies outlined below.

The same propulsion technologies that our interplanetary descendants will use to move efficiently through the solar system, such as nuclear fission or more likely nuclear fusion, perhaps using fuels laced with antimatter to make fusion more efficient, could take us to the nearest stars if we are willing to wait long enough. Multi-generation starships able to reach the nearest stars in centuries could be built with moderate technological improvements.

But, even with efficient fusion propulsion, a faster starship would have to carry too much fuel to be viable. Antimatter propulsion [Forward 1995], much more efficient than nuclear fusion propulsion, would permit accelerating a starship to a significant fraction of the speed of light (ten percent or more), and decelerating to stop at the destination, with realistic amounts of fuel on board.

Antimatter powered starships could reach the nearest stars in decades instead of centuries. We are unable to produce the required amounts of antimatter at this moment, but things could change.

The Bussard ramjet wouldn't have to carry fuel. Instead, it would collect fuel on the fly by using magnetic fields to scoop the atoms present in interstellar space. As noted by Carl Sagan [Sagan 1966], the concept is very elegant and could be a game changer. While the original idea of Robert Bussard is out of fashion at this moment, some promising variants are worth pursuing.

Light sails pushed by focused laser or microwave beams in the solar system [Drexler 1986, Forward 1995, Loeb 2021] would also permit accelerating interstellar starships without fuel on board. The fuel would remain at home, so to speak, and there are apparently viable ways to decelerate and stop at the destination. Combining the technologies used in Bussard ramjets and light sails, we can think of magnetic sails [Zubrin 2019] that use a magnetic field as a sail pushed by the solar wind or a particle beam.

Instead of beaming radiation or particles to propel a sail, we can think of shooting fuel pellets at a starship. Or even better, we can think of pre-positioning fuel pellets along the path of a starship to provide an acceleration runway in the solar system.

While requiring improvements in technology, these interstellar propulsion systems sit comfortably within mainstream consensus science. Other more imaginative propulsion systems could rely on reactionless drives based on the hypothetical possibilities of pushing against the quantum vacuum, tapping the "zero point energy" [Forward 1995, Tajmar 2003] that is present in empty space, or rowing through empty space by inducing variations of inertial mass as suggested by James Woodward [Cramer 2013, Woodward 2013]. According to John Cramer, Woodward's theory "could have important general relativity implications" for the warp drives and wormholes discussed below.

Future spacefarers could travel among the stars in style. Really in style. Faster than light.

After Einstein, consensus physics has ruled out the possibility of moving through space faster than light. But space itself can move faster than light, as in the early "inflationary" phase of the expansion of the universe. Empty space is not a nothing, but a something that can be shaped by matter and energy consistently with Einstein's general relativity. The Alcubierre warp drive [Cramer 2013] would allow "to 'inflate' the space in back of a starship... and to 'deflate' the space in front of the starship" [Forward 1995], transporting the starship like a surfer carried by a faster than light wave of space.

Traversable wormholes [Thorne 1994, 2014, Cramer 2013] that would allow, so to speak, to go around space rather than through it, are also entirely consistent with Einstein's general relativity. A traversable wormhole would provide a faster than light path between its two ends separated in space. It's worth noting that, under certain conditions, suitably arranged wormholes would also allow time travel to the past.

Using warp drives or traversable wormholes for spaceflight would require the ability to engineer exotic forms of matter and energy that, while apparently consistent with physics, are entirely beyond us at this moment. But not necessarily beyond a very advanced civilization "whose activities are limited only by the laws of physics" [Thorne 1994]. Hopefully, we will become a very advanced civilization ourselves.

As an aside, let me go back to the ideological packages discussed in [Chapter 5]. Today, many scientists and science fans seem committed to an all-out war on imagination in the name of excessive caution. They have knee-jerk reactions when they hear about highly imaginative ideas such as faster than light warp drives or time travel, and accuse highly imaginative colleagues of being "crazy science fiction physicists" [Thorne 1994].

I guess this is a residual of past struggles to free science and society from some toxic effects of religious dogmatism, but these "scientific justice warriors" throw out the baby with the dirty bath water. Their warped logic seems to be that scientific optimism and speculative frontier science open the door to anti-scientific thinking and threaten to bring back magical thinking or even (God forbid) religion.

But Avi Loeb rightly warns [Loeb 2021] that:

> "The tenuous threads connecting humanity's Earth-bound civilization as it exists today, and the promise of humanity's possible interstellar civilization as it might exist tomorrow, will not be upheld by exercising conservative caution."

The 100 Year Starship project [Benford 2013] strives to make human interstellar spaceflight a reality within the next 100 years.

I'll now outline two interstellar mission projects that have been proposed and could be enabled by technology developments that can be expected to materialize soon. One wants to send ultra-light miniaturized probes to the nearest star and return data after only a few decades. Another wants to send a probe, which could one day be followed by people, to a very interesting destination at the edge of interstellar space.

The Breakthrough Starshot project [Loeb 2021] wants to send the first robotic probes to the nearest star, Alpha Centauri, in only a few decades and with only a few decades of travel time.

If things go according to plans (things seldom do, but optimism feels good), some readers will be alive when the first data and images come back from the Alpha Centauri system.

Starshot is the first interstellar probe project with sufficient funding for a thorough feasibility study. In fact, initial $100 million funding was provided by Russian billionaire Yuri Milner. At the Starshot announcement event in April 2016, Stephen Hawking said:

> "The limit that confronts us now is the great void between us and the stars. But now we can transcend it. With light beams, light sails and the lightest spacecraft ever built, we can launch a mission to Alpha Centauri within a generation. Today, we commit to this next great leap into the cosmos. Because we are human. And our nature is to fly."

The key elements of Starshot are based on technology either already available or likely to be attainable in the near future under reasonable assumptions, and the Alpha Centauri mission is expected to require a budget comparable to the largest current scientific experiments: the total funding needed would be of the order of $5-10 billion, which seems ambitious but possible.

The star probe will be a highly miniaturized system on a chip, propelled by a light sail built with advanced nano-engineered materials. The probe and its light sail, both weighing only a few grams, will be pushed by light beams from high power lasers, accelerated to 20 percent of the speed of light, and reach Alpha Centauri in two decades. There's no room for a deceleration system, so the mission will be a high speed fly-by with the goal of returning data and images.

Among the many Starshot system design challenges, data return is expected to be one of the hardest, because there is little room for a communication system able to send data back to Earth.

A first stepping stone on the road to the stars could be a mission to establish a gravitational lens observatory in deep interstellar space, at about three light days from the Earth. This is the distance of the gravitational focus of the Sun, where gravitational lensing provides a huge amplification of signals from the opposite direction (the Sun must be between the observatory and the target). Carl Sagan explains [Sagan 1994]:

> "If you are free to roam an imaginary spherical shell at the appropriate focal distance and centered on the Sun, you are free to explore the Universe in stupendous magnification, to peer at it with unprecedented clarity, to eavesdrop on the radio signals of distant civilizations, if any, and to glimpse the earliest events in the history of the Universe. Alternatively, the lens could be used the other way, to amplify a very modest signal of ours so it could be heard over immense distances."

Perhaps we need a gravitational lens router in place if we want to join the galactic internet of distant civilizations. Claudio Maccone has proposed [Maccone 2009, 2012] to get started with a first exploratory mission, called FOCAL, to the gravitational focus of the Sun.

It's worth noting that a gravitational lens observatory pointed at Alpha Centauri could permit receiving the faint signals sent back from the small, low power transmitters of Starshot probes. Therefore, a FOCAL mission could be an important element of Starshot.

In *"Aurora"* [Robinson 2015], Kim Stanley Robinson imagines the first human interstellar mission, launched in 2545. The story starts in a large multi-generation starship with more than two thousand human starfarers en route to Tau Ceti at one tenth of the speed of light, then moves to Aurora, an Earth-like moon in the Tau Ceti system.

"Aurora" is a fascinating interstellar adventure, with plausible pic-

tures of interstellar propulsion and a quantum computer slowly waking up to sentience. But the hard problem is biology.

Robinson argues that a multi-generation starship would be too small to include a viable ecosystem able to support life for hundreds of years, and would inevitably fail. Another problem is that "If it's alive it's going to be poisonous." A nasty alien micro-organism found on Aurora, from which there is no defense, threatens to exterminate the starfarers and eventually forces them to head back to the Earth.

I find Robinson's scenario excessively grim to the point of inconsistency. For example, the starfarers have powerful nanotechnology that can analyze and engineer matter at molecular scales. I find it strange that they are unable to use their nanotechnology to stabilize life support on the starship and fight the alien pathogen on Aurora.

In *"The Next 500 Years"* [Mason 2021], inspired by recent advances in synthetic biology and genetic engineering, Christopher Mason suggests that we will eventually engineer human DNA to enable future generations to settle other planets in the solar system and around other stars. Genetically modified humans could be adapted to live on the surface of other planets or in engineered habitats and interstellar starships.

Another possibility is to upload human minds to advanced computers in a starship. Computers don't require air, water, food, or medical care, and withstand acceleration and radiation much better than organic bodies. Therefore, the size and weight of the starship can be dramatically reduced.

Future technologies [Kurzweil 2005, Bainbridge 2007, Prisco 2020] like nanotechnology, neurotechnology and mind uploading (personality transfer) could be the ultimate enablers of interstellar spaceflight and human expansion to the stars. The minds, personalities, memories, and feelings of the crew members would be moved to computing circuitry on board.

I'm not saying that today's digital technologies can process human consciousness and "run" uploaded human minds. I don't think this is the case. But I'm persuaded that future technologies will reproduce the yet poorly understood features that enable biological brains to process consciousness. New kinds of computing circuitry, perhaps advanced quantum chips, will be able to run uploaded human minds.

Once the interstellar destination is reached, the uploaded minds of the crew members would be moved to suitably designed nanoprinted bodies. Perhaps one of the first tasks of the crew could be the construction of receivers for other spacefarers, who will then join the interstellar outpost traveling as radiation and light beams.

With these technologies, the galactic network of gravitational lens routers outlined above would also be a galactic transportation network. In the science fictional universe of *"Manifold: Space"* [Baxter 2000], Stephen Baxter explores this possibility.

If strong Artificial Intelligence (AI) [Kurzweil 2005, Bainbridge 2007, Prisco 2020] is developed, perhaps way smarter than humans [Lovelock 2019], why should we bother to send uploaded human minds to the stars? Isn't AI good enough?

One answer is that we want human minds among the stars. Another answer is that future people will be human/AI hybrids [Kurzweil 2005], blended so tightly that it will be impossible to tell which is which.

Of course we'll meet alien civilizations among the stars [Sagan 1966, Freitas 1979]. We haven't found confirmed evidence of extraterrestrial life yet, but I'm persuaded that the universe is structured in such a way that conscious intelligent life (perhaps much different from life as we know it [Freitas 1979, Vidal 2014]) inevitably emerges and thrives.

Interstellar futures have been imagined by many science fiction writers, with plenty of detail. Unfortunately, most vintage science fiction doesn't feel credible today because it portrays people like us and societies like ours, only with futuristic technologies. This is, I think, far too naive.

Greg Egan envisions less naive interstellar futures in [Egan 2002, 2008]. Orion's Arm, a collaborative project to imagine plausible interstellar futures, has produced a really spectacular science fictional universe. Besides published collections of short stories e.g. [Orion's Arm 2014] and a novel [Bowers 2012], the project maintains a sprawling website at orionsarm dot com. The website includes an "*Encyclopaedia Galactica*" with thousands of entries and counting.

The Orion's Arm project was started in 2000 by Alan Kazlev and Donna Hirsekorn, who "wanted stories set in a future which might really happen" [Orion's Arm 2014]. I was involved in a project to create a virtual world based on Orion's Arm. This project eventually stalled, but I hope others will continue it.

The Orion's Arm universe, set ten thousand years from now, spans thousands of light years with countless worlds and space habitats. People range from "near baseline" to heavily modified humans with all sorts of body plans and embedded technology, including superhumans with extremely advanced augmentations and AI subsystems. Most people are virtual beings living as pure software.

Engineered wormholes are used for long distance interstellar hauls, but wormhole physics doesn't allow using wormholes for time travel. A few alien civilizations have been found, but none advanced as humans. There are, however, clues that suggest very advanced alien civilizations that existed in the past.

Directed superhuman evolution has produced vast God-like beings with mega brains, internally connected by instantaneous wormhole links, which span star systems and light years. Only these beings can understand and create some extremely advanced technologies used in Orion's Arm.

This short outline doesn't even begin to do justice to the vast complexity of Orion's Arm. Visit orionsarm dot com for much more. Or even better, participate in the project. I can promise that Orion's Arm will give you awesome dreams and a burning enthusiasm for our interstellar future.

9 - COSMIC ENGINEERS

One day, future scientists will visit black holes, experiment with extreme physics, and make awesome new discoveries.

To reach a faraway black hole, our descendants will need to master science much more advanced than ours. They'll need to understand things like quantum gravity and deep quantum vacuum physics pretty well.

Perhaps they'll smugly think that they already know everything there is to be known. But I think new, unexpected experimental results in the extreme conditions near a black hole will shatter their certainty and force them to develop new science. The history of science shows that new previously unfeasible experiments lead to unexpected, paradigm breaking scientific developments.

I am persuaded that the search for more accurate scientific models of reality will never end. I don't think a finite Theory of Everything is there to be found. On the contrary, I think physical reality has an infinitely deep structure [Prisco 2020] that will never stop showing unpredictable twists and turns.

Shakespeare said it better:

> "There are more things in heaven and earth, Horatio, than are dreamt of in our philosophy."

There will always be little or big cracks in current theories, better theories lurking behind the cracks, then new cracks in the new theories, and so forth. Our models of reality are thin, but reality itself is thick. Physical reality will always, I think, be infinitely more complex than our models, and full of unexpected surprises.

The Italian futurists [Chapter 6] would agree. Riccardo Campa argues [Campa 2012] that their artistic provocations were informed by a worldview close to Henri Bergson's philosophy of becoming, quite different from the philosophy of being embraced by many thinkers including Laplace, which was the conceptual paradigm of 19th century science.

Laplace's universe is pristine, aseptic, mechanical, reducible, deterministic, and reversible. But Bergson's universe is messy, dirty, living, whole, nondeterministic, and irreversible. Campa explains that, according to Bergson, creative evolution is present everywhere in nature. Matter is alive, flowing, mysterious, spiritual. The philosophy of Bergson, and of the Italian futurists, opens the door to a magic conception of the world.

In modern times Alfred North Whitehead, Ilya Prigogine, Erich Jantsch [Jantsch 1980] and many scientists, especially biologists but also chemists and physicists, have embraced parts of Bergson's philosophy. Becoming is more fundamental than being, and comes before being. Complexity, life, consciousness and intelligence are favored by principles embedded in natural laws, which operate at all scales.

There's "intelligent directionality in the evolution of the universe" [Wolfe 2015], and intelligent life is "very much part of the universe's DNA."

The writings of the Italian futurists are full of visionary ideas such as humanity remaking itself by merging with technology, overcoming all limits, and ascending to the stars. In Campa's words (my translation):

> "Marinetti and the futurists set themselves objectives like, no less, 'to challenge the stars', 'to ascend to the sky', 'to reconstruct the universe', 'to create the mechanical man with interchangeable parts'..."

It's worth noting that more and more contemporary scientists and engineers are saying exactly the same things [Kurzweil 2005]. According to Christopher Mason [Mason 2021], humanity will remake itself with synthetic biology and genetic engineering, and ascend to the stars. We will "engineer at a genetic, cellular, planetary, and interstellar scale" and eventually "reengineer the universe itself" to ensure that life continues to thrive indefinitely. Mason argues that this is our "duty to the universe and to life itself."

Campa explains that, according to the Italian futurists (my translation):

> "The goal of science is not solving practical problems, but shedding light on the unknown. But here comes an unexpected twist… The futurists startle everyone with a statement on the edge of paradox: they affirm that knowledge is an intermediate goal, since the ultimate goal of science is… mystery!"

Contemporary science suggests that the universe is a quite mysterious place indeed, with hidden dimensions, instant connections between different places and times and whatnot [Prisco 2020]. And I think this is good. Like the Italian futurists, I am persuaded that good science keeps us dreaming of "magic" hidden realities, and driven to develop new science and technology to access and act upon these realities.

The Russian cosmists [Chapter 6] would agree. But they were even more visionary and daring. According to cosmist mystic Nikolai Fedorov, the mentor of spaceflight pioneer Konstantin Tsiolkovsky, future scientists and engineers will be able to resurrect the dead, bringing back to life every person who ever lived.

Fedorov thought that future engineers will resurrect the dead by finding the atoms that formed their bodies and putting them back in place. Fedorov's technological resurrection theory reflects 19th-century models of the universe and seems naive today, but new theories based on contemporary science have been proposed

[Prisco 2020].

Of course, our new theories will probably seem equally naive tomorrow. Our descendants will devise better theories, then preliminary experiments, then even better theories, and so forth. Fedorov must be credited for the idea of technological resurrection. Following Fedorov, future engineers will scan the fabric of reality to find the dead, and bring them back to life.

Eventually, we will become cosmic engineers and do wonderful big things. Like, you know, remaking the universe and resurrecting the dead.

Out there among the stars, we'll gradually learn how to do these things. Perhaps we'll also meet benevolent teachers, advanced civilizations that are much farther along the way.

The excellent book "*Indistinguishable From Magic*" [Forward 1995], which I have cited many times, borrows its title from Arthur Clarke's third law:

> "Any sufficiently advanced technology is indistinguishable from magic."

An example of magic technology in the sense of Clarke's third law would be engineering and using wormholes [Chapter 8] not only for faster than light spaceflight, but also for time travel to the past.

Contrary to what is often said, time travel to the past doesn't necessarily introduce consistency paradoxes, which can be avoided in two ways. First, we can think that the past and the future adjust to each other self-consistently, without paradoxes [Thorne 1994, Forward 1995]. Second, we can think of parallel timelines [Prisco 2020].

I think a general paradox free solution could be a combination of the two. That is, the universe tries to keep everything unfolding self-consistently in one timeline, but brings a new timeline into play when self-consistency is impossible to achieve in the original one.

The combination of time travel to the past and mind uploading implies the possibility of technological resurrection [Prisco 2020]. Too bad the wormholes in the collaborative science fictional universe of Orion's Arm [Chapter 8] don't allow time travel, which would open the door to great stories of people like us resurrected in the Orion's Arm universe after death.

Another example of magic technology in the sense of Clarke's third law is the fate of my favorite character in *"The Expanse"* [Chapter 7]. Miller, the noir detective who falls in love with a ghost and then dies to save the Earth, eventually becomes a ghost himself. The ultra-powerful "protomolecule" nanotechnology left behind by God-like aliens builds a replica of Miller's mind in some exotic otherworldly substrate, and beams Miller's mind to Jim Holden now and then. How this works is not spelled out in detail, but there are hints that there must be weird quantum shenanigans at work.

Stanley Kubrick's film *"2001: A Space Odyssey"* doesn't explain much of what happens behind the scene, because Kubrick wanted "to remove as much verbal explication as possible in favor of purely visual and sonic cues" [Benson 2018]. After seeing the film with my mother I asked her what the monolith was. She said that perhaps (as I was beginning to suspect) it was God.

Clarke explains more in the novel developed jointly with the film [Clarke 1968], envisioning ultra advanced civilizations in the universe:

> "In their ceaseless experimenting, they had learned to store knowledge in the structure of space itself, and to preserve their thoughts for eternity in frozen lattices of light. They could become creatures of radiation, free at last from the tyranny of matter. Now they were lords of the galaxy, and beyond the reach of time. They could rove at will among the stars, and sink like a subtle mist through the very interstices

of space."

My interpretation of these words of Clarke is that advanced life forms in the universe could eventually migrate to the fundamental fabric of space and time, quantum vacuum fields or whatever lies beneath [Prisco 2020].

Robert Freitas [Freitas 1979] and Clément Vidal [Vidal 2014] speculate on exotic life forms based on nuclear interactions, which could exist in high energy environments such as neutron stars. Even more exotic life based on gravitational fields, the fabric of apparently empty space, could exist around or inside black holes.

Intelligent minds could run on matter substrates much faster and more powerful than biological brains. Such minds could emerge spontaneously or be engineered by other intelligent beings, which then might choose to port themselves to new substrates.

Another leap of imagination brings us to intelligent life based directly on quantum vacuum physics. I think this is what Clarke had in mind.

In a 1968 interview [Norden 1968], Kubrick doesn't say how to interpret the film, which must be left to the viewer, but elaborates on his conviction "that one can construct an intriguing scientific definition of God." Kubrick evokes God-like beings that emerge from matter "transformed into beings of pure energy and spirit."

> "These entities might be in telepathic communication throughout the cosmos and thus be aware of everything that occurs, tapping every intelligent mind as effortlessly as we switch on the radio; they might not be limited by the speed of light and their presence could penetrate to the farthest corners of the universe; they might possess complete mastery over matter and energy; and in their final evolutionary stage, they might develop into an integrated collective immortal consciousness. They would be incomprehensible to

us except as gods; and if the tendrils of their consciousness ever brushed men's minds, it is only the hand of God we could grasp as an explanation."

In the novel, Clarke compares them with "something which, long ago, men had called spirit."

"And if there was anything beyond that, its name could only be God."

So what about God?

In 1948 Olaf Stapledon gave a talk titled "Interplanetary Man?" [Stapledon 1948] at the British Interplanetary Society. The talk was organized by Clarke.

Our conception of time itself "is now turning out to be very incoherent and superficial," said Stapledon. "Perhaps (who can say) from the point of view of eternity the end of the cosmos is also its source and its temporal beginning."

"Perhaps the ultimate flower is also the primal seed from which all sprang. Perhaps the final result of the cosmical process is the attainment of full cosmical consciousness, and yet (in some very queer way) what is attained in the end is also, from another point of view, the origin of all things. So to speak, God, who created all things in the beginning, is himself created by all things in the end."

The passage above is one of the best formulations of the idea that God emerges from the physical universe and comes to full being and power at the end of time. But the God that exists at the end of time is present and acts in the universe at all earlier times via self-consistent causal loops in time [Prisco 2020].

The idea that God emerges (Emerged? Will emerge?) in the far future of the physical universe is also suggested by the works of scientists like Fred Hoyle and Frank Tipler, and theologians like Pierre Teilhard de Chardin and Wolfhart Pannenberg [Prisco

2020]. According to Frank White [White 2018] we will create, or become part of, the universal mind. Martine Rothblatt [Chapter 6] envisions a cosmic deity in the making, which will eventually "encompass the universe, thus becoming omniscient, omnipotent and omnificent" [Terasem 2012].

Clarke reportedly suggested that:

> "It may be that our role on this planet is not to worship God but to create him."

This futurist, cosmist religion is my personal religion. In my group "Turing Church" we discuss these and related ideas. Look for us on the internet, and you'll find us at turingchurch dot net, other websites, and social media. In my previous book *"Tales of the Turing Church"* [Prisco 2020] I have described my religion in much more detail.

My religion is not based on the "supernatural." On the contrary, I think we live in one natural universe that we'll understand better (though perhaps never entirely) in the future. But in our natural universe there are things that are so much beyond our understanding that "supernatural" seems the only honest description at this moment. If physical reality has an infinitely deep structure as suggested above, this will always be the case.

I realize that my religion seems inspired by science and science fiction instead of traditional religion (actually, it is). But I'm saying that God exists, and we'll live again. Isn't this exactly what traditional religions teach? Therefore, I emphasize the parallels between my religion and traditional religions more than the differences.

Allow me to go back, once again, to the ideological packages discussed in [Chapter 5, 8]. The same ideological package that supports environmentalism and social justice, but condemns spaceflight, nuclear energy, and imagination, also condemns religion.

In the words of Kubrick [Norden 1968], "there is a certain element of the lumpen literati that is so dogmatically atheist and materialist and earth-bound that it finds the grandeur of space and the myriad mysteries of cosmic intelligence anathema."

I have said many times that I support environmentalism and social justice, but also spaceflight, nuclear energy, and imagination. I'll add that I am persuaded that religion can be a force for good, and I would welcome the religious renaissance discussed in [Chapter 6].

In a visionary essay [Bainbridge 2009], William Sims Bainbridge argues that the "creation of a galactic civilization may depend upon the emergence of a galactic religion capable of motivating society for the centuries required to accomplish that great project."

So far, I have said nothing about the personal nature of God. But it seems to me that the ultimate being that eventually emerges from the physical universe is not less than personal, but at least personal, and likely much more than personal.

Following traditional religions, I think God is compassionate. We are (or should be) compassionate, we should strive to become much more so, and I think we will. If God emerges from what humanity will become, it seems plausible that God is at least compassionate, and likely much more than compassionate.

So we can think that God watches us here and now, perhaps with loving care, and perhaps God answers our prayers now and then.

God will recruit (has recruited?) us as helpers and apprentice cosmic engineers. Armed with "divine" science and technology, our descendants will resurrect the dead and remake the universe.

If humanity ends before that point, other civilizations in the universe will play this role. But I like to think that our descendants will be there and participate.

Our present homework is to start hiking the road to the stars.

First, we must become an interplanetary civilization here in the solar system as soon as possible, to avoid cultural senility and catastrophic events that could destroy a civilization confined to this planet. Then, we'll go to the stars and beyond.

This is our hope, our great expectation, our duty, and our destiny.

REFERENCES

[Allaby 1984] Michael Allaby, James Lovelock. *The Greening of Mars*. Warner Books, 1984.

[Asimov 2018] Isaac Asimov, Frederik Pohl. *Our Angry Earth: A Ticking Ecological Bomb*. Tor Books, 2018 (first published in 1991).

[Autino 2020] Adriano Autino. *A greater world is possible: The expansion of civilization beyond the limits of our home planet is the moral issue of our time*. Adriano Autino, 2020.

[Azurmendi 1992] Joxe Azurmendi. *El hombre cooperativo: Pensamiento de Arizmendiarrieta*. Azatza, 1992.

[Bainbridge 1976] William Sims Bainbridge. *The Spaceflight Revolution: A Sociological Study*. Wiley, 1976.

[Bainbridge 2007] William Sims Bainbridge. *Nanoconvergence: The Unity of Nanoscience, Biotechnology, Information Technology and Cognitive Science*. Prentice Hall, 2007.

[Bainbridge 2009] William Sims Bainbridge. Religion for a Galactic Civilization 2.0. *IEET*, 2009.

[Bainbridge 2015] William Sims Bainbridge. *The Meaning and Value of Spaceflight: Public Perceptions*. Springer, 2015.

[Bassani 2018] Giulia Bassani. *Ad Martem 12*. Giulia Bassani, 2018.

[Baxter 2000]. Stephen Baxter. *Manifold: Space*. Voyager, 2000.

[Bell 2017] Tom Bell. *Your Next Government? From the Nation State to Stateless Nations*. Cambridge University Press, 2017.

[Benford 2013] Gregory Benford, James Benford. *Starship Century:*

Toward the Grandest Horizon. Microwave Sciences, 2013.

[Benson 2018] Michael Benson. *Space Odyssey: Stanley Kubrick, Arthur C. Clarke, and the Making of a Masterpiece*. Simon & Schuster, 2018.

[Bowers 2012] Steve Bowers. *Betrayals*. Orion's Arm Universe Project, 2012.

[Brand 1977] Stewart Brand. *Space Colonies*. Whole Earth Catalog, 1977.

[Brand 2010]. Stewart Brand. *Whole Earth Discipline*. Penguin, 2010.

[Brown 2016] Kerry Brown. *CEO, China: The Rise of Xi Jinping*. I.B. Tauris & Co., 2016.

[Butler 1993] Octavia Butler. *Parable of the Sower*. Four Walls Eight Windows, 1993.

[Butler 1998] Octavia Butler. *Parable of the Talents*. Seven Stories Press, 1998.

[Calhoun 1962] John Calhoun. Population Density and Social Pathology. *Scientific American*, 1962.

[Campa 2012] Riccardo Campa. *Trattato di Filosofia Futurista*. Avanguardia 21, 2012.

[Canavan 2016] Gerry Canavan. *Octavia E. Butler*. University of Illinois Press, 2016.

[Chua 2018] Amy Chua. *Political Tribes: Group Instinct and the Fate of Nations*. Penguin, 2018.

[Clarke 1967] Arthur Clarke. *The Coming of the Space Age*. Meredith Press, 1967.

[Clarke 1968] Arthur Clarke. *2001: A Space Odyssey*. New American Library, 1968.

[Clarke 1979] Arthur Clarke. *The Fountains of Paradise.* Gollancz, 1979.

[Clarke 1994] Arthur Clarke. *The Snows of Olympus: A Garden on Mars.* Gollancz, 1994.

[Cockell 2015a] Charles Cockell (Ed.). *The Meaning of Liberty Beyond Earth.* Springer, 2015.

[Cockell 2015b] Charles Cockell (Ed.). *Human Governance Beyond Earth: Implications for Freedom.* Springer, 2015.

[Corey 2011] James Corey. *Leviathan Wakes.* Orbit, 2011.

[Corey 2012] James Corey. *Caliban's War.* Orbit, 2012.

[Corey 2013] James Corey. *Abaddon's Gate.* Orbit, 2013.

[Corey 2014] James Corey. *Cibola Burn.* Orbit, 2014.

[Corey 2015] James Corey. *Nemesis Games.* Orbit, 2015.

[Corey 2016] James Corey. *Babylon's Ashes.* Orbit, 2016.

[Corey 2017] James Corey. *Persepolis Rising.* Orbit, 2017.

[Corey 2019] James Corey. *Tiamat's Wrath.* Orbit, 2019.

[Corey 2021] James Corey. *Leviathan Falls.* Orbit, expected 2021.

[Cramer 2013]. John Cramer. Exotic Technologies for Interstellar Travel. In [Benford 2013].

[Crichton 2004] Michael Crichton. *State of Fear.* HarperCollins, 2004.

[Criswell 2002] David Criswell. Solar Power via the Moon. *The Industrial Physicist*, 2002.

[Cruddas 2020] Sarah Cruddas. *Look Up: Our story with the stars.* HQ, 2020.

[Danesi 2018] Marcel Danesi. *Of Cigarettes, High Heels, and Other*

Interesting Things. Palgrave Macmillan, 2018.

[Daum 2019] Meghan Daum. *The Problem with Everything: My Journey Through the New Culture Wars*. Gallery Books, 2019.

[Davenport 2018] Christian Davenport. *The Space Barons: Elon Musk, Jeff Bezos, and the Quest to Colonize the Cosmos*. PublicAffairs, 2018.

[David 2019] Leonard David. *Moon Rush: The New Space Race*. National Geographic, 2019.

[Del Toro 2019] Guillermo del Toro. The Most Radical and Rebellious Choice You Can Make Is to Be Optimistic. *Time*, 2019.

[Defense 2020] Office of the Secretary of Defense. *Military and Security Developments Involving the People's Republic of China 2020: Annual Report to Congress*. Office of the Secretary of Defense, 2020.

[Deudney 2020] Daniel Deudney. *Dark Skies: Space Expansionism, Planetary Geopolitics, and the Ends of Humanity*. Oxford University Press, 2020.

[Douthat 2020] Ross Douthat. *The Decadent Society: How We Became the Victims of Our Own Success*. Avid Reader Press, 2020.

[Drexler 1986] Eric Drexler. *Engines of Creation: The Coming Era of Nanotechnology*. Doubleday, 1986.

[Dyson 1979]. Freeman Dyson. *Disturbing the Universe*. Harper & Row, 1979.

[Egan 2002] Greg Egan. *Schild's Ladder*. Gollancz, 2002.

[Egan 2008] Greg Egan. *Incandescence*. Gollancz, 2008.

[Eisenstein 2011] Charles Eisenstein. *Sacred Economics: Money, Gift, and Society in the Age of Transition*. Evolver Editions, 2011.

[Fallaci 2014] Oriana Fallaci. *Se il sole muore*. Rizzoli, 2014 (first published in 1965).
English translation: Oriana Fallaci. *If the Sun Dies*. Atheneum,

1966.

[Ferreira 2018] Becky Ferreira. We Asked 105 Experts What Scares and Inspires Them Most About the Future. *Vice*, 2018.

[Forward 1995] Robert Forward. *Indistinguishable From Magic*. Baen, 1995.

[Freitas 1979] Robert Freitas. *Xenology: An Introduction to the Scientific Study of Extraterrestrial Life, Intelligence, and Civilization*. Xenology Research Institute, 1979.

[Fuller 1969] Buckminster Fuller. *Operating Manual for Spaceship Earth*. Southern Illinois University Press, 1969.

[Gates 2021] Bill Gates. *How to Avoid a Climate Disaster: The Solutions We Have and the Breakthroughs We Need*. Alfred A. Knopf, 2021.

[Gertz 2019] Bill Gertz. *Deceiving the Sky: Inside Communist China's Drive for Global Supremacy*. Encounter Books, 2019.

[Gilster 2004] Paul Gilster. *Centauri Dreams: Imagining and Planning Interstellar Exploration*. Springer, 2004.

[Goswami 2020] Namrata Goswami and Peter Garretson. *Scramble for the Skies: The Great Power Competition to Control the Resources of Outer Space*. Lexington Books, 2020.

[Gribbin 2009] John and Mary Gribbin. *He Knew He Was Right: The Irrepressible Life of James Lovelock and Gaia*. Penguin, 2009.

[Groys 2018] Boris Groys (Ed.). *Russian Cosmism*. MIT Press, 2018.

[Healey 2019] Robin Healey. *Italian Literature since 1900 in English Translation: An Annotated Bibliography, 1929-2016*. University of Toronto Press, 2019.

[Jantsch 1980] Erich Jantsch. *The Self-Organizing Universe: Scientific and Human Implications of the Emerging Paradigm of Evolution*. Pergamon Press, 1980.

[Kaczynski 2010] Theodore Kaczynski. *Technological Slavery: The Collected Writings of Theodore J. Kaczynski, a.k.a. The Unabomber.* Feral House, 2010.

[Kaku 2018] Michio Kaku. *The Future of Humanity: Terraforming Mars, Interstellar Travel, Immortality, and Our Destiny Beyond.* Penguin, 2018.

[Kilday 2018] Bill Kilday. *Never Lost Again: The Google Mapping Revolution That Sparked New Industries and Augmented Our Reality.* Harper Business, 2018.

[Kissinger 2011] Henry Kissinger. *On China.* Penguin, 2011.

[Kornuta 2019] David Kornuta et al. Commercial lunar propellant architecture: A collaborative study of lunar propellant production. *REACH - Reviews in Human Space Exploration,* 2019.

[Kurzweil 2005] Ray Kurzweil. *The Singularity Is Near: When Humans Transcend Biology.* Viking Press, 2005.

[Lakoff 2014] George Lakoff.*Don't Think of an Elephant: Know Your Values and Frame the Debate.* Chelsea Green Publishing, 2014.

[Lakoff 2016] George Lakoff. *Moral Politics: How Liberals and Conservatives Think.* University of Chicago Press, 2016.

[Landis 2013] Geoffrey Landis. The Nuclear Rocket: Workhouse of the Solar System. In [Benford 2013].

[Launius 2019] Roger Launius. *Apollo's Legacy: Perspectives on the Moon Landings.* Smithsonian Books, 2019.

[Lewis 2014] John Lewis. *Asteroid Mining 101: Wealth for the New Space Economy.* Deep Space Industries, 2014.

[Liu 2013] Liu Cixin. *The Wandering Earth.* Beijing Guomi, 2013,

[Liu 2013a] Liu Cixin. Sun of China. In [Liu 2013].

[Liu 2014] Liu Cixin. *The Three-Body Problem.* Tor Books, 2014.

[Liu 2015] Liu Mingfu. *The China Dream: Great Power Thinking and Strategic Posture in the Post-American Era*. CN Times Books, 2015.

[Liu 2015a] Liu Cixin. *The Dark Forest*. Tor Books, 2015.

[Liu 2016] Ken Liu (Ed.). *Invisible Planets: Contemporary Chinese Science Fiction in Translation*. Tor Books, 2016.

[Liu 2016a] Liu Cixin. The worst of all possible universes and the best of all possible earths: Three-Body and Chinese science fiction. In [Liu 2016].

[Liu 2016b] Liu Cixin. *Death's End*. Tor Books, 2016.

[Liu 2109] Ken Liu (Ed.). *Broken Stars: Contemporary Chinese Science Fiction in Translation*. Tor Books, 2019.

[Loeb 2021] Avi Loeb. *Extraterrestrial: The First Sign of Intelligent Life Beyond Earth*. Houghton Mifflin Harcourt, 2021.

[Lovelock 2000] James Lovelock. *Gaia: A New Look at Life on Earth*. Oxford University Press, 2000 (first published in 1979).

[Lovelock 2019] James Lovelock. *Novacene: The Coming Age of Hyperintelligence*. MIT Press, 2019.

[Maccone 2009] Claudio Maccone. *Deep Space Flight and Communications: Exploiting the Sun as a Gravitational Lens*. Springer, 2009.

[Maccone 2012] Claudio Maccone. *Mathematical SETI: Statistics, Signal Processing, Space Missions*. Springer, 2012.

[Mahbubani 2020] Kishore Mahbubani. *Has China Won?: The Chinese Challenge to American Primacy*. PublicAffairs, 2020.

[Mailer 1970] Norman Mailer. *Of a Fire on the Moon*. Little Brown, 1970.

[Mankins 2014] John Mankins. *The Case for Space Solar Power*. Virginia Edition Publishing, 2014.

[Mason 2021] Christopher Mason. *The Next 500 Years: Engineering*

Life to Reach New Worlds. MIT Press, 2021.

[McCracken 2012] Garry McCracken, Peter Stott. *Fusion: The Energy of the Universe*. Academic Press, 2012.

[Millis 2009] Marc Millis, Eric Davis (Eds.). *Frontiers of Propulsion Science*. American Institute of Aeronautics and Astronautics, 2009.

[Newell 2019] Catherine Newell. *Destined for the Stars: Faith, the Future, and America's Final Frontier*. University of Pittsburgh Press, 2019.

[Newitz 2013] Annalee Newitz. Stop pretending we aren't living in the Space Age. *io9*, 2013.

[O'Neill 2013] Gerard O'Neill. *The High Frontier: Human Colonies in Space*. Space Studies Institute, 2013 (first published in 1977).

[Naam 2013] Ramez Naam. *The Infinite Resource: The Power of Ideas on a Finite Planet*. University Press of New England, 2013.

[Nevala-Lee 2018] Alec Nevala-Lee. *Astounding: John W. Campbell, Isaac Asimov, Robert A. Heinlein, L. Ron Hubbard, and the Golden Age of Science Fiction*. Dey Street Books, 2018.

[Norden 1968] Eric Norden. Playboy Interview: Stanley Kubrick. *Playboy*, 1968. Republished by *Playboy* as *Stanley Kubrick: The Playboy Interview*, 2012.

[Orion's Arm 2014] Orion's Arm Universe Project. *After Tranquility*. Orion's Arm Universe Project, 2014.

[Parisi 2019] Jason Parisi, Justin Ball. *The Future Of Fusion Energy*. World Scientific, 2019.

[Prisco 2009] Giulio Prisco. A Virtual World Space Agency. *Futures*, 2009.

[Prisco 2017] Giulio Prisco. Why We Need a Decentralized Autonomous Space Agency. *Vice*, 2017.

[Prisco 2020] Giulio Prisco. *Tales of the Turing Church: Hacking religion, enlightening science, awakening technology*. Giulio Prisco, 2020.

[Rainey 2009] Lawrence Rainey, Christine Poggi, Laura Wittman (Eds.). *Futurism: An Anthology*. Yale University Press, 2009.

[Robinson 2012] Kim Stanley Robinson. *2312*. Orbit, 2012.

[Robinson 2015] Kim Stanley Robinson. *Aurora*. Orbit, 2015.

[Robinson 2018] Kim Stanley Robinson. *Red Moon*. Orbit, 2018.

[Robinson 2020] Kim Stanley Robinson. *The Ministry for the Future*. Orbit, 2020.

[Sagan 1966] Carl Sagan, Iosif Shklovsky. *Intelligent life in the universe*. Holden-Day, 1966.

[Sagan 1994] Carl Sagan. *Pale Blue Dot: A Vision of the Human Future in Space*. Random House, 1994.

[Sapolsky 2017] Robert Sapolsky. *Behave: The Biology of Humans at Our Best and Worst*. Penguin, 2017.

[Schmidle 2021] Nicholas Schmidle. *Test Gods: Virgin Galactic and the Making of a Modern Astronaut*. Henry Holt and Company, 2021.

[Scott-Heron 2000] Gil Scott-Heron. *Now and Then*. Canongate, 2000.

[Shetterly 2016] Margot Lee Shetterly. *Hidden Figures: The American Dream and the Untold Story of the Black Women Who Helped Win the Space Race*. Harper Collins, 2016.

[Schmitt 2006] Harrison Schmitt. *Return to the Moon: Exploration, Enterprise, and Energy in the Human Settlement of Space*. Copernicus, 2006.

[Spudis 2016] Paul Spudis. *The Value of the Moon: How to Explore, Live, and Prosper in Space Using the Moon's Resources*. Smithsonian

Books, 2016.

[Stapledon 1937] Olaf Stapledon. *Star Maker*. Methuen, 1937.

[Stapledon 1948] Olaf Stapledon. Interplanetary Man? *Journal of the British Interplanetary Society*, 1948. Republished in [Clarke 1967].

[Strittmatter 2019] Kai Strittmatter. *We have been harmonised: Life in China's surveillance state*. Old Street Publishing, 2019.

[Suarez 2019] Daniel Suarez. *Delta-v*. Dutton, 2019.

[Tajmar 2003] Martin Tajmar. *Advanced Space Propulsion Systems*. Springer, 2003.

[Terasem 2012] Terasem Movement. *Truths of Terasem*. Terasem Movement, 2012.

[Thiel 2020] Peter Thiel. Back to the Future. *First Things*, 2020.

[Thorne 1994] Kip Thorne. *Black Holes & Time Warps: Einstein's Outrageous Legacy*. W. W. Norton, 1994.

[Thorne 2014] Kip Thorne. *The Science of Interstellar*. W. W. Norton, 2014.

[Turner 2011] Frederick Jackson Turner. *The Frontier in American History*. Barnes & Noble, 2011.

[Vidal 2014] Clément Vidal. *The beginning and the end: the meaning of life in a cosmological perspective*. Springer, 2014.

[Wiener 2018] Anna Wiener. The Complicated Legacy of Stewart Brand's "Whole Earth Catalog." *The New Yorker*, 2018.

[White 2018] Frank White. *The Cosma Hypothesis: Implications of the Overview Effect*. Multiverse Publishing, 2018.

[White 2021] Frank White. *The Overview Effect: Space Exploration and Human Evolution*. Multiverse Publishing, 2021. First published by Houghton-Mifflin, 1987.

[Wiener 2018] Anna Wiener. *The Complicated Legacy of Stewart Brand's Whole Earth Catalog*. The New Yorker, 2018.

[Wolfe 1979] Tom Wolfe. *The Right Stuff*. Farrar, Straus and Giroux, 1979.

[Wolfe 2015] Steven Wolfe. *The Obligation: A Journey to Discover Human Purpose on Earth and in the Cosmos*. 2015.

[Woodward 2013] James Woodward. *Making Starships and Stargates. The Science of Interstellar Transport and Absurdly Benign Wormholes*. Springer, 2013.

[Young 2012] George Young. *The Russian Cosmists: The Esoteric Futurism of Nikolai Fedorov and His Followers*. Oxford University Press, 2012.

[Zubrin 1997] Robert Zubrin. *The Case for Mars: The Plan to Settle the Red Planet and Why We Must*. Touchstone, 1997.

[Zubrin 2019] Robert Zubrin. *The Case for Space: How the Revolution in Spaceflight Opens Up a Future of Limitless Possibility*. Prometheus Books, 2019.

www.ingramcontent.com/pod-product-compliance
Lightning Source LLC
Chambersburg PA
CBHW070424220526
45466CB00004B/1539